The Enchantment of the Long-haired Rat

Tim Bonyhady is one of Australia's foremost environmental historians. His books include *The Colonial Earth*, which won both the New South Wales Premier's Prize for Australian History and the Queensland Premier's Prize for History, and his prize-winning family memoir, *Good Living Street*.

The Enchantment *of the* Long-haired Rat

A Rodent History of Australia

TIM BONYHADY

TEXT PUBLISHING MELBOURNE AUSTRALIA

textpublishing.com.au

The Text Publishing Company
Swann House, 22 William Street, Melbourne Victoria 3000, Australia

Published by The Text Publishing Company, 2019.

Cover design by W. H. Chong.
Cover illustration by H. C. Richter, *Mus longipilis*, 1854, from John Gould's *Mammals of Australia*.
Page design by Jessica Horrocks.
Typeset in Granjon by J&M Typesetting.
Map by Simon Barnard.
Index by Mary Russell.

Printed and bound in Australia by Griffin Press, part of Ovato, an accredited ISO/NZS 14001:2004 Environmental Management System printer.

ISBN: 9781925773934 (paperback)
ISBN: 9781925774689 (ebook)

A catalogue record for this book is available from the National Library of Australia.

This book is printed on paper certified against the Forest Stewardship Council® Standards. Griffin Press holds FSC chain-of-custody certification SGS-COC-005088. FSC promotes environmentally responsible, socially beneficial and economically viable management of the world's forests.

For Nicole Moore
of Balranald

Contents

PART III

Decline

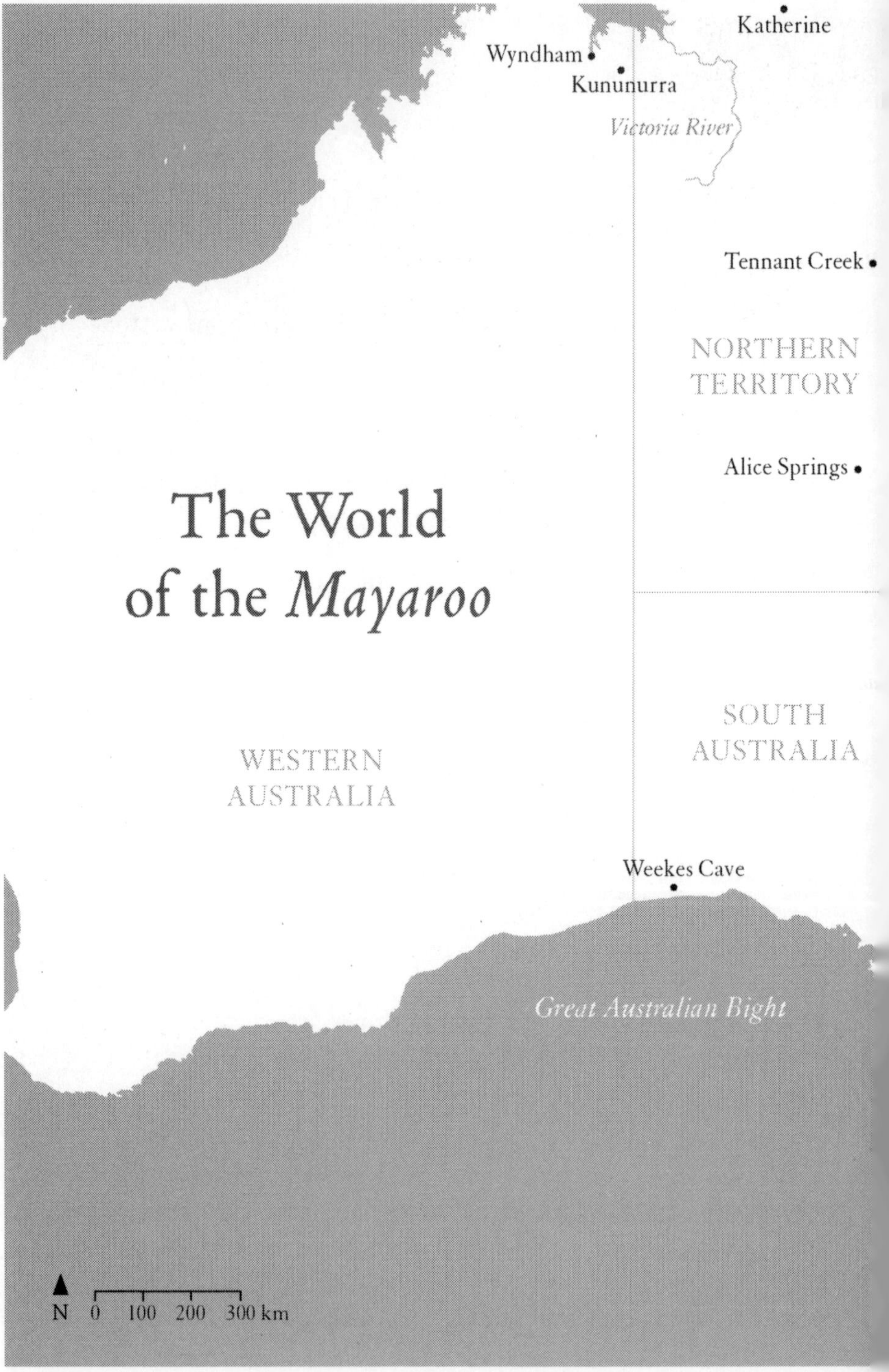

The World of the *Mayaroo*

Katherine

Wyndham

Kununurra

Victoria River

Tennant Creek

NORTHERN TERRITORY

Alice Springs

WESTERN AUSTRALIA

SOUTH AUSTRALIA

Weekes Cave

Great Australian Bight

N 0 100 200 300 km

Gulf of Carpentaria
Edward
Islands
Burketown
Flinders River
Townsville
Barkly Tableland
Camooweal
Charters Towers
Hughenden
Winton
Boulia
Diamantina River
Longreach
Barcaldine
Barcoo River
Bulloo River
Birdsville
QUEENSLAND
Cooper Creek
Paroo River
Innamincka
Thargomindah
Lake Hope
Lake Killalpaninna
Strangway Springs
Milparinka
Marree
Bourke
Walgett
Tilpa
Wilcannia
Silverton
Liverpool Plains
Port Augusta
Darling River
Teetulpa
Menindee
NEW SOUTH WALES
Ivanhoe
Lachlan River
Wentworth
Hay
Murrumbidgee River
Port Lincoln
Balranald
Swan Hill
Murray River
ACT
Kerang
VICTORIA

INTRODUCTION

They talked of myriads, legions, swarms and armies—of visitations, invasions and plagues. Estimates of their numbers ran to not just hundreds or thousands but millions. Where passenger pigeons were once so abundant in North America that they blocked out the sun, Australia's long-haired rats left no trace of whatever preceded them. By one account, they obliterated wheel and horse tracks for miles along a main road as they 'paddled down the sandy soil like a flock of sheep'. According to another, all tracks through a pastoral station 'were erased each night by the passing swarms, as if the surface of the soil had been swept by a broom'.

These episodes were all the more remarkable because the rats were usually few and rarely to be seen. Their plagues were even more extraordinary because, as the rats multiplied and spread, they attracted large numbers of dingoes, snakes and birds of prey which also multiplied as they fed on the rats. Then, when the plagues ended, these predators found other food, dispersed or died of starvation or disease. One was the letter-winged kite, a raptor unique to Australia, particularly dependent on the rats. The earliest European drawing of it, produced at the new settlement at Port Jackson, is the only surviving sign that in the 1790s there was a spectacular increase and then collapse in the number of rats in the far inland, forcing the kites to quit their usual terrain for the coast in search of other prey.

Aboriginal people across much of Australia have delighted in the long-haired rat as an easy source of abundant food. Some groups took it as their totem, vital to their social organisation and integral to their cosmology. The Diyari people east of Lake Eyre were probably far from alone in staging ceremonies to draw the rats towards them and increase their numbers. While Lutheran missionaries in the late nineteenth and early twentieth centuries sought to force the Diyari to abandon their traditions and embrace Christianity, the Diyari maintained these ceremonies to enchant the rat.

Colonists, on the whole, brought with them the deep-seated European fear and loathing of rats and ate the Australian ones only if starving, if they ate them at all. The colonists also were attacked by the long-haired rat in a way that Aborigines were not, as the Europeans' supplies and possessions provided a novel lure. Far from entering their cosmology, the rat became part of settler demonology for over a century, decried as 'pests' and 'vermin' and much else. Yet settlers occasionally recognised the rat's plagues as 'a most extraordinary natural occurrence' and identified it as 'fantastic' and 'wonderful'—rare, perhaps even unprecedented words for a rat, resonating with the Aboriginal desire to enchant it.

The rainfall in the Australian interior is notoriously variable—perhaps more so than that of any other continent's arid regions. When there are big rains and floods, prompting many plants to grow prodigiously, animals respond to this abundance of food in diverse ways. Some species increase in number only modestly. Others are more responsive. A relatively small group goes through cycles of boom and bust, multiplying in spectacular fashion when conditions are good, only to collapse when the land dries out and there is little to eat, leaving remnants surviving in areas scientists have come to call 'refuges'. Until the late twentieth century, the long-haired rat was one of the great irruptive species of the planet, exceptional for the extent of the terrain it occupied and the duration of its plagues. One of seven Australian species of *Rattus*, where the Americas have none, it has been the native rat of greatest human interest and consequence since at least colonial times. Aboriginal people gave it many names including *artoka*, *gootanga* and *yimala*. Western science first knew it as *Mus longipilis*, the long-haired mouse. Now it is *Rattus villosissimus* or the hairy rat, commonly called the long-haired rat.

Its ancestors may have arrived from New Guinea about 1.4 million years ago, with *Rattus villosissimus* diverging from another species within Australia about 500,000 years ago, but physical evidence of the long-haired rat before European colonisation is limited. Its fossilised bones have been found in just one place. There are further traces of it in old pellets of barn owls, found in caves and overhangs, which often contain bones sufficiently intact to allow their identification. The first European records of the rat, from the 1840s, reveal it then reached further south and, possibly, further east than it has since. But these records are thin because colonists were yet to occupy the rat's prime domains and, when they did, settlers often did not write about or depict what they witnessed. It was not until an unusually long period of huge rains and vast floods, extending from 1885 into 1888, that a big irruption became the stuff of extensive contemporaneous recording—revealing how, at the plague's peak, much of the continent became an immense rattery.

This plague occurred before dams and irrigation transformed the flow of water across the country; before cattle and camels destroyed many of the long-haired rat's prime habitats; and before human-induced climate change. But the colonisers had shaped what occurred. Almost a century after Europeans invaded the continent, they had put an end to almost all traditional Indigenous management of land. Their sheep had compacted the soil and eaten out native grasses. Rabbits were beginning to reach much of the terrain of the rat and compete with it for food and burrows. Cats were preying on the rats in ever more areas—spreading not only because they were abandoned or strayed but also because some pastoralists bought and released as many as they could in the vain hope they might control the rabbits devastating the land. The Europeans' stores of food and leather goods also attracted the rat, influencing where it congregated and how long it stayed.

The result was something fantastical, and not just because it saw the first encounter of the rat and the rabbit, with the rat heading south and the rabbit heading north, and colonists firmly on the side of the rat, hoping it might stop the rabbit. The encounter of these rodents—as the rabbit and the rat were both categorised at the time—occurred in a landscape which, from the settlers' perspective, was largely out of control, exciting the colonial imagination. Talk of sightings of old creatures from the Dreaming of Aborigines and of new hybrid species proliferated. So did accounts of cupidity and fraud by the settlers themselves, as they competed for the spoils of the land and colonial society fractured along class lines in the face of an environmental and economic crisis.

When species become extinct, historians and scientists sometimes try to reconstruct how they once lived. 'Forensic ornithology', the American writer Jonathan Rosen characterised such work, in relation to the passenger pigeon. When species still exist, there is rarely the same interest, for all the species may have declined. The rat is one example. Its irruptions have never been as immense as colonists witnessed in the 'rat years' between 1885 and 1888. Nor will it irrupt again on the same scale—environmental destruction by settler Australians has been too profound, even before climate change.

Recovering the lost world of the long-haired rat involves many challenges. There are big gaps—long silences—when the rat was not irrupting or no one bothered to write about it. When records exist, there is the problem of interpreting observations literally made in the dark by people who often had little or no scientific knowledge. That problem is heightened when accounts are inconsistent or when it is unclear which species—or whether more than one species—was involved. All this becomes even harder to untangle when the focus is an object of hatred and hyperbole.

The black or English rat—the first to spread from Asia, perhaps reaching Spain in the second century BCE—was feared and loathed in Europe because it was so destructive. The species dubbed the brown or Norway rat, despite its also coming out of Asia, followed in the sixteenth century and, when it became dominant in the eighteenth and nineteenth centuries, the fear and loathing of rats was largely transferred to it. The German naturalist Peter Pallas declared the Norway rat 'the most foul, the most ferocious, the most pernicious' of all species. Charles Dickens' *Household Words* considered it 'the emblem of mystery, murder and rapine'. The British naturalist J.G. Millais reckoned the Norway 'the best-hated animal in Europe', only for that hatred to fix again on the English rat once it was accepted that its fleas caused the pandemic of bubonic plague that started in the 1890s.

A key characteristic of the English and Norway rats is that they are commensal—they depend directly on humans and are primarily urban. The long-haired rat is not commensal, for all its contact with people during irruptions. It also is not a threat to human health. Still, the exotic rats have shaped the response to the Australian ones. The very term 'rat' has been and remains an obstacle to sympathy. 'Plague rat', as the Australian animal has often been dubbed, is even more alienating. Colonists arrived with a lexicon for rat writing, a wealth of rat clichés and an array of literary references from *Aesop's Fables* to the Pied Piper and beyond, which shaped how the long-haired rat was described. Hyperbole inflated some estimates of the number of rats and reports of what they did. Little wonder that most modern scientists have disregarded old accounts, though they may have done that anyway because of their assumption that only refereed journal articles are worth considering.

I first learned about the long-haired rat while writing a book about the explorers Burke and Wills, who encountered the rat when crossing the continent in 1860–61. I was struck by the wealth of evidence of the rat in old lists of Aboriginal words, letters, diaries, newspapers, paintings, prints, photographs, magazines, scientific publications and books. This abundance of material was all the more striking given the increasingly thin modern science based on either laboratory studies or fieldwork confined to small areas, often not central to major irruptions at their peak. I thought that by exploring this great array of sources—artistic, linguistic and journalistic; scientific, documentary and literary; anthropological and legal—and pursuing the rat across time, I might be able to show how the rat was once a very different creature from what it is now: found in very different country, interacting with different animals, and with a very different place in the lives of people, Indigenous and non-Indigenous. I thought the rat could be an entry point into another world—in fact, several other worlds—as well as a subject in its own right.

The climate—its fluctuations between vast floods and intense droughts, and ever-increasing change—is pivotal to the rat's story. The phenomena we now know as La Niña, causing immense rains, and El Niño, causing extended dry periods, loom large. Because the great spikes in the rat's numbers are typically associated with La Niñas, these events are integral to this history. Because collapses in the rat's numbers are primarily associated with El Niños, these episodes are also crucial. But the rat also reveals that, for all the significance of La Niñas and El Niños, the consequences of these events are often highly complex. Many La Niñas have not led to irruptions while some of the rat's plagues have begun in El Niño years and others continued through them.

The book's terrain extends from the Ord River in Western Australia to the Nullarbor Plain and Port Lincoln on the South Australian coast, from Townsville and Charters Towers in northern Queensland through New South Wales to northern Victoria. Here is a story of the rivers of western Queensland, which in El Niño years are often just a succession of isolated billabongs but in La Niña years become vast watercourses that typically break their banks and flow through the Channel Country, occasionally reaching

Lake Eyre in South Australia. Here, too, is a story of pastoral stations conventionally measured according to their number of stock and of outback towns measured by their numbers of hotels. While the Bulloo and the Diamantina were Australia's great rat rivers, Winton in western Queensland became Australia's first and foremost rat town.

Other species loom large, especially the long-haired rat's predators, native and introduced—most notably, the letter-winged kite, inland taipan, dingo, fox and cat. The rat's story also involves many other creatures that have occupied the same terrain, sometimes in prodigious numbers. One is the small nocturnal marsupial, the bilby, now in great jeopardy. Another is the flock bronzewing pigeon, the closest Australian counterpart to the passenger pigeon, which once also filled the skies—a Queensland bushman reckoned it 'the most wonderful thing in bird life he had ever known'.

Not least, *The Enchantment of the Long-haired Rat* is about people. It examines the rat's place in the diet, language, geography and cosmology of Aboriginal groups—especially the Diyari, east of Lake Eyre, for whom the rat is the *mayaroo*. It explores how Australia's settlers struggled to understand the rat's sudden appearance in great numbers, its spread across the landscape and its rapid disappearance. It reveals the difficulties of overcoming inherited, deep-seated fear of a species and of classifying and learning about it, especially when this species is nocturnal. It is a story about ignorance of the past, the richness of amateur observations and the limitations of professional science; a story about how hard it is to learn about any species, how long it takes and how little we know. And it is a story about the stories told about rats.

German Australians loom large, since they were at the

forefront of colonial science and art and missionary activity. Chinese Australians also play a big role as the market gardeners and cooks of so many places in the outback. But the colonist central to this book—the figure who more than any other recurs across its chapters—was born in Tasmania of English parentage, moved as a boy to Gippsland in Victoria, then spent almost all his adult life in the far west of New South Wales. A station manager-cum-rabbit inspector, often 'hard up' and episodically 'one of the great unemployed', of scant public interest in his lifetime and largely ignored ever since, Kenric Harold Bennett appears conventionally in surviving photographs as another bearded man in a three-piece suit.

He was in many ways a typical frontiersman. As a station manager, Bennett played a part in the colonists' dispossession of the country's Indigenous owners and in the conflict between the settlers over whether the land should be the preserve of wealthy 'squatters' or distributed more equitably to 'selectors' of small landholdings. In maximising financial returns from sheep and cattle, he contributed to the devastation of native species. He and his station hands were once held up by bushrangers. Because of his work, he was absent from his family for most of the time. With infant mortality rife among the colonists, the first three of his six children died before they turned five, with two of the children dying within a day.

Bennett was also a rarity—one of a small class of 'bush naturalists' without formal training, who, especially in his case, observed Australia's rapidly changing environment with great perspicacity. For years, he had no means of disseminating what he learned. Then he began gaining access to public forums where he could record and interpret what he, often alone among colonists, had been studying and collecting. His eagerness to ingratiate himself with professional scientists caused him to rob Aboriginal graves, though he recognised that doing so was desecration. His delight in the natural environment, omnivorous interests, deep curiosity and keen awareness of the settlers' destructiveness remain exemplary. More than any other European, he observed and wrote about the long-haired rat and other species that shared its terrain when the rat was only beginning to be affected by colonisation.

A rodent history of Australia has a great choice of subjects. Of the 270 or so species of mammal unique to the continent, there are about sixty non-marsupial rats and mice, almost all paid scant attention. More than eighty years ago, one of the finest zoologists of the Australian interior, H.H. Finlayson, declared in *The Red Centre*: 'There is a widespread idea amongst Australians that all the "native" animals are marsupials. But nowhere in the country can that opinion be proved more fallacious than in the Centre, where true indigenous rodents vastly predominate numerically over the marsupials, and even in the point of species and genera, make a brave show in the list.' More recently, biologist Tim Low wrote, 'Australia is blessed with a bounty of remarkable rodents.' This book tells the story of one as wondrous in its way as the koala, kangaroo and platypus.

Part I

The Land of the *Mayaroo*

CHAPTER 1

A Word from the Barngarla

A word—not even a phrase, let alone a sentence—provides a start. Its spelling is a little odd, its meaning more of an issue. Its place of origin is most interesting when one of the great questions about the long-haired rat is where it ranged before colonists began transforming and reducing its terrain. The recorder of the word was Clamor Schürmann of the Lutheran Missionary Society of Dresden in his second book about the Aborigines of South Australia, published six years after his arrival in Adelaide. *Mai erri*, as the twenty-nine-year-old pastor rendered it in 1844, is both a puzzle and a clue.

It appears in a list of words of the people whom Schürmann thought of as Parnkalla, now called Barngarla. Having already combined with a fellow German missionary to list 2000 words of the Kaurna people, whose lands included the Adelaide plains,

Schürmann recorded more than 3000 words of the Barngarla, because the Lutherans' philosophy was to proselytise in the language of those whom they sought to convert. While Schürmann probably encountered the Barngarla primarily around Port Lincoln, near the western tip of Spencer Gulf, where he took up a government post as deputy-protector of Aborigines in 1840, the Barngarla's vast semi-arid country stretched west across much of the Eyre Peninsula and north beyond Port Augusta.

'A species of mouse' was how Schürmann translated *mai erri*. Not 'a species of mouse or rat', 'a common rat', 'a species of rat', 'a burrowing animal; rat', or a 'kangaroo rat', as Schürmann translated other words. By implication, the *mai erri* was small. A reader, considering only Schürmann's book, would never think it was the long-haired rat, which is the biggest of the Australian species of *Rattus*, though nothing like the size of some other Australian rodents such as the water rat. If *mai erri* were a word unique to the Barngarla, there would be no basis for suggesting Schürmann might have been mistaken, since Barngarla is now one of those Aboriginal languages being revived but still barely used, with no speakers to elucidate its meaning.

Mai erri is, however, almost certainly one of those words used by Aboriginal groups across much of South Australia that Europeans would write down in diverse ways due to the difficulties of representing Aboriginal sounds using the Latin alphabet. While later Lutheran missionaries to the Diyari people east of Lake Eyre tried *majaru*, *maijaru*, *maiaru* and *mairuh*, other colonists in this region opted for *miaroo*. A word list for the Karangura people, north of the Diyari, included *mi-arroo*. A journalist reported that the neighbouring Arabana people called them *myroo*. A zoologist recorded *mai-ai* for the Ngamini people on the Diamantina River. A linguist settled on *mayadu* for the Diyari. Another linguist decided on *maiurru* for the Adnyamathanha people of the Flinders Ranges and *maiarri* for the Barngarla. After seeking more linguistic advice, a team of scientists called for Aboriginal and non-Aboriginal people alike to use *mayaroo*.

The word's meaning has been the subject of more confusion. A few accounts suggest a marsupial, thereby excluding the long-haired rat. Edward B. Sanger, a geologist in South Australia's far north at the start of the 1880s, mistakenly identified it as the swamp rat, predominantly a south-eastern species, found close to the coast. Zoologist H. H. Finlayson recorded in the 1930s that the Wangkangurru people used *miaroo* for the long-haired rat, but heard that 'in an earlier usage the word had a more general application to many, if not all kinds of rats'. Yet several accounts reveal that, as Aboriginal people differentiated between species due to the importance of rats to their diet and culture, Aborigines used this word for the long-haired rat. They did so, linguist Dorothy Tunbridge has elucidated in the most illuminating discussion of *mai erri* and its many variants, because of the meaning of the word's two components. Its opening *mai* means vegetable food in many languages. Its final *erri* means 'all' in one language, consistent with how the rat is primarily herbivorous, and means 'grey'—the rat's prime colour—in at least one other language.

All this indicates that Clamor Schürmann erred in his translation, a common problem with nineteenth-century word lists, even those recorded by colonists with considerable linguistic skills such as Schürmann. The *mai erri* was a rat, not a mouse. While one of the great scholars of Aboriginal languages, Luise Hercus, warned about the temptation to ask too much of colonial lists of Aboriginal words because there are so few other sources for many languages, Schürmann's list indicates that the terrain of the long-haired rat overlapped with that of the Barngarla people. *Mai erri* is a linguistic sign that the rat could at least sometimes be found on the Eyre Peninsula, perhaps even extending as far south as Port Lincoln.

CHAPTER 2

The Blunders of Science

Explorers' journals provide more evidence. In September 1847, a party led by Edmund Kennedy, a surveyor employed by the New South Wales Government, was on Cooper Creek, 600 miles north-east of Port Lincoln. With the clay soils of the plains almost bare and deeply cracked due to lack of rain, Kennedy recorded that his horses fell 'up to their knees at every step'. His deputy, Alfred Turner, who reckoned this country 'defied description'—while of course trying to describe it—noted that these cracks offered 'first rate accommodation' to 'great numbers of rats and snakes'. When the party's natural history collector, Thomas Wall, tried to catch them, the snakes eluded him but he caught several of the rats.

Charles Sturt was among the few Australian explorers who tried to bring back some of his catches alive. His first attempt, after departing Adelaide in search of an inland sea in 1844, was with pig-footed bandicoots—small marsupials with hoof-like toenails resembling the cloven feet of pigs. These bandicoots, now extinct, appear to have lived on a mix of insects and plants. Sturt thought they preferred the meat and offal of birds, which he fed them until they died. He tried next with hopping mice which seemingly did better. They were 'thriving beautifully on oats' until a 'dreadful day' when one of Sturt's men fastened 'a tarpaulin down over them...by which means they were smothered'.

Most explorers sought to return with dead specimens, but often failed. As Thomas Mitchell, the Surveyor-General of New South Wales, approached the junction of the Murray and Darling Rivers on his first major expedition in 1835, his men killed an animal that Mitchell identified as a kind of rat with ears resembling those of a small, wild rabbit. Eager for this creature to become the stuff of science, Mitchell ordered it be preserved, but one of his men fed it to his dog. Thomas Wall was more successful in 1847 with the rats from Cooper Creek. Perhaps he preserved a pair whole in alcohol. Perhaps, given he was a taxidermist, he skinned them on the spot. Either way, Wall got at least one to Sydney where, in 1848, he announced that, while 'very like in external appearance to the common English rat', known now as *Rattus rattus*, it was 'entirely new'.

Just a few years before, this rat would have gone to London and stayed there—partly because there was no colonial institution able to care for it, partly because of the power wielded by British scientists eager for new specimens. Although a museum was established in Sydney in 1829, it had six homes in its first fifteen years, all cramped and inadequate. When the German scientist Ludwig Leichhardt arrived in 1842, he quipped: 'A museum did once exist in Sydney, but it is now in such a state of confusion, that I was not able to catch a glimpse of it, in fact I believe nobody knows at present where it is to be found.'

Leichhardt was pivotal to this shambolic institution—known from 1836 as the Australian Museum—becoming a prime repository of new material. In doing so, he challenged the established system of colonial observation and collection of species before description, classification, publication and theorising in Europe. Leichhardt's reasons were partly self-interested. Because he expected to remain in Australia, he wanted to be able to work with a strong, local collection. But Leichhardt also looked to facilitate the research of other scientists in Australia. The German Leichhardt—disparaged as a 'foreigner' by his British contemporaries in Sydney—was the first great Australian scientific nationalist.

Leichhardt decided to put Australia before Europe in 1844 when organising his first major expedition—a private venture without government funding—which would see him become the first colonist to explore the far north-east of the continent. Although Leichhardt had little money, he resolved to give away the specimens he found, and give Sydney priority over Berlin, Paris and London. If he returned with one specimen of a species, he would donate it to the Australian Museum. If he returned with more than one, the best would stay in Sydney while European institutions received the others.

Leichhardt's generosity encouraged the museum to aspire to a collection that would induce 'learned professors...to come among us, and kindle in all classes the love of the sciences'. To achieve this goal, the museum began compelling explorers who were government officials, unlike Leichhardt, to treat their collections as public property, as their contracts of employment required. The museum started with Thomas Mitchell, who retained much of the material from his last expedition in 1845. Mitchell promised to comply once he had laid this material 'before persons of science in England for classification'. By 1847, he had delivered 850 specimens.

When Edmund Kennedy went to the Cooper in 1847, and the government failed to include a natural history collector in his party, Kennedy employed Thomas Wall out of his own pocket. As a result, Kennedy owned what Wall collected. But when the government reimbursed Kennedy for Wall's wages, the rat was among 376 specimens acquired by the Australian Museum. Its curator, William Sheridan Wall, Thomas's brother, immediately affirmed that the rat was new and, most likely, briefly displayed it in the room in Sydney's Darlinghurst courthouse which was then the museum's temporary home. Another natural history collector, Frederick Strange, also identified the rat as new in a letter to one of the pivotal figures of global science in London, John Gould, who was the pre-eminent classifier of Australian birds and mammals.

Gould's prime business was publishing lavish bird books, which owed much of their success to their coloured plates, created by his wife Elizabeth, with scant acknowledgment, from 1830 until her death in 1841. Having begun with two volumes of birds of the Himalayas, Gould followed with multi-volume ventures devoted to Europe, Australia, Asia, Great Britain and New Guinea, and produced monographs on the toucan and the partridges of America and the hummingbird. *The Birds of Australia* occupied Gould from 1840 and ran to thirty-six parts, seven volumes and six hundred plates, followed by an eighty-one-plate supplement, as he aspired to publish a folio-sized, hand-coloured lithograph of every Australian bird with an accompanying page of text. Having begun *The Mammals of Australia* in 1845 with similar ambitions, Gould wanted to include Thomas Wall's rat.

Gould had visited Australia from 1838 until 1840, but had no intention of returning to examine specimens in Sydney. Instead, he expected them to come to him. Having secured a loan of Leichhardt's finds from the Australian Museum with Leichhardt's agreement, he similarly sought those collected by Wall. Dr George Bennett, a Sydney medical practitioner, who was one of the colony's leading naturalists and a member of the museum's governing committee as well as the Australian agent for Gould's lavish books, advised Gould that some reciprocity would yield what he wanted. If he provided the skins of big mammals, such as lions and leopards, which the museum was eager to display, it would send the specimens Gould required. When Gould refused, the rat remained in Sydney awaiting description and classification.

Description was simple since contemporary science accepted a brief account of the animal's external appearance. But confirming whether the rat was unknown to Europeans, as the Wall brothers and Frederick Strange maintained, was much more of a challenge involving comparison with other specimens in European collections. So was the selection of the binomial Latin name required by the Linnean system of classification. The animal had to be placed in the correct genus, then the species needed to be given a name not used before, and that required access to a substantial scientific library, which Sydney lacked.

Publication was yet another challenge for colonists because of the expectation that the account be in a scientific journal, museum catalogue or appendix to an explorer's travels. In the late 1840s, there was only one Australian scientific periodical—published in Hobart by the Royal Society of Tasmania and largely confined to its proceedings. Kennedy's diary might have formed the basis of a book, but in 1848, on his next expedition, to Cape York, he was speared and killed by Aborigines, along with nine of his men, including Thomas Wall who was accompanying him again.

Convention was against local classification. In the first half of the nineteenth century, colonial naturalists almost never named species. Since the rules of zoological nomenclature, developed in England in the early 1840s, specified that species were what 'competent' naturalists said they were, the colonists implicitly accepted they were incompetent. While Thomas Wall's rat was in Sydney, William Sheridan Wall did nothing with it.

Gould finally prevailed in 1852 without giving the Australian Museum anything. While rats are notorious for crossing oceans as unwanted passengers on ships and canoes, the one from Cooper Creek became much desired, deliberate cargo when the museum sent it to Gould in London, along with seventeen other specimens from Kennedy's first expedition. In 1854, Gould published it in the sixth part of his *Mammals of Australia*, expressing his gratitude to the museum's trustees, while making plain his regret that he had to return this 'remarkable', 'unique' specimen to Sydney.

Gould's lithograph of the rat was by H. C. Richter, whom Gould employed for forty years, following Elizabeth's death, to turn thousands of Gould's rudimentary drawings of a bird's or animal's shape, size and position into relatively accurate but generally mundane lithographs. Richter depicted Wall's rat twice—once from the front, once from the side—his standard mode whether he had access to just one specimen or several, so he could reveal more of a species' appearance. When zoologist H. H. Finlayson assessed Richter's rendering, he declared it 'good' while criticising its general shape as 'insufficiently rakish', its head as 'too chubby and too small', its hair as 'more erect than it normally is' and its tail as 'much too light in colour'. In other words, not all that good.

The text—probably written by Gould, assisted by his secretary Charles Prince—included a few measurements for the rat, which Gould likened in size to the English or black rat, and a brief description of its appearance. Otherwise, the text reworked information from William Sheridan Wall in Sydney, who had reported that, after catching several of the rats, his brother dissected some, curious about their diet. Thomas found their stomachs contained 'a fleshy mass', prompting him to conclude 'they preyed on each other' since he saw no other animals there, apart from the snakes. Gould provided a gloss, correctly suggesting the rat was primarily a herbivore, with its cannibalism 'only temporary, probably caused by the entire absence at the time of the seeds and other vegetable substances suitable to its economy'.

William Sheridan Wall's identification of the source of the rat was misleading. Kennedy's instructions in 1847 were to follow a river already encountered by Thomas Mitchell, who thought it would flow north-west through the interior and prove to be the Victoria, which has its mouth in northern Australia on the Timor Sea. Instead, the river followed by Kennedy flowed south-west and, given the opportunity to name it, Kennedy embraced its local Aboriginal name, the Barcoo. It was a tributary of Cooper Creek, which Charles Sturt had named in 1845, and Kennedy's party was on the Cooper when Thomas Wall collected the rat. But because William Sheridan Wall described the expedition as 'to the Victoria River', so did Gould, fueling longstanding geographic confusion.

Gould was capable of great analysis. When Charles Darwin returned in 1836 from his voyage on the *Beagle*, and wanted his specimens classified by England's foremost scientists, his bird collection was examined by Gould, whose close study of the beaks of the Galapagos finches proved pivotal to Darwin's theory of natural selection. But when it came to Wall's rat, Gould made one mistake after another. Lack of expertise was one cause. *The Mammals of Australia* was Gould's only major publication not devoted to birds—an experiment he would not repeat. Because his *Mammals* drew about half as many subscribers as his *Birds of Australia*, it earned him much less than any of his big bird books. But Gould was also careless, if not lazy. His repeated blunders undermined the standard justification for sending colonial specimens to Europe—that scientists there were best equipped to classify them.

Gould bungled when he decided the rat from the Cooper was a new species—a mistake that bedeviled nineteenth-century science. A common reason was the dispersal of specimens across countries, if not continents, which made comparison difficult if not impossible. But the two other specimens that Gould needed to examine were, like him, in London. Even more remarkably, Gould himself had collected these rats on his visit to Australia, and done so before Wall found his example on the Cooper. At least in this case, Gould might have made do with specimens immediately available to him, without pressing the Australian Museum for its holdings or lamenting having to return them.

Gould's rats came from the Liverpool Plains, their most easterly recorded location in the continent's south-east. When he visited over the summer of 1839–1840, the country was exceptionally bountiful due to big rains, enabling him to secure 'a glorious haul' of specimens. While Gould presented some of this material at the monthly meetings of the Zoological Society of London after he returned to England, the rats were not his priority. Instead, he included them in a collection of specimens bought by the British Museum in 1841, which its zoologists would come to regard as one of the museum's great acquisitions. In 1843, John Edward

Gray, its zoological keeper, identified the rats collected by Gould on the Liverpool Plains as *Pseudomys*, a catch-all for a miscellany of rodents from Australia and New Guinea. Most likely, Gould did not bother to inspect these *Pseudomys greyi* because the Wall brothers and Frederick Strange thought the rat from the Australian Museum was new. Possibly Gould looked but was blind to their resemblance to Wall's specimen.

Gould identified Wall's specimen as a *Mus*, the Latin term for both mice and rats, which Carl Linnaeus, the architect of modern taxonomy, applied to the English and Norway rats and the house mouse. In naming Wall's specimen, Gould fixed on its guard hairs protecting the rest of its fur. Because of the 'great length' of these 'numerous black hairs interspersed along the back' of the rat, Gould dubbed it *longipilis*.

The Linnean system is predicated on each species receiving its own distinctive scientific name, but that often did not happen. When Gould applied *Mus longipilis* to Wall's rat, this name had already been used by the curator of the Zoological Society of London, George Waterhouse, for a Chilean rat collected by Charles Darwin while voyaging on the *Beagle*. So had *Pseudomys greyi* when John Edward Gray used it.

The standard explanation, articulated by the American historian of science Harriet Ritvo, is that not even the most vigilant naturalist could 'locate and examine all potentially relevant reports—buried as they might be in the proceedings of obscure societies, published in foreign languages'. Gould's choice of *Mus longipilis* exemplifies how the practice of naming was even more deeply flawed. When George Waterhouse first used *Mus longipilis*, he did so not in a Chilean journal published in Spanish but at a meeting of the Zoological Society in London with Gould present. After Waterhouse displayed and described this rat in 1837, Gould displayed and described five finches collected by Darwin on the *Beagle*. Waterhouse's account of the Chilean *Mus longipilis* appeared in the *Proceedings of the Zoological Society of London* immediately before Gould's description of the finches.

Even if Gould and his secretary Charles Prince did not think to search these *Proceedings* for earlier uses of *Mus longipilis*, they could have checked Waterhouse's catalogue of mammals held by the Zoological Society of London which first appeared in 1838. Alternatively, they might have checked the second part of Darwin's five-part *Zoology of the Voyage of H.M.S. Beagle* published in 1839, which contained a more detailed account of Waterhouse's *Mus longipilis* and an illustration of it. Like the Zoological Society's *Proceedings* for 1837, the *Zoology of the Voyage of H.M.S. Beagle* would have been on Gould's shelves, since he contributed to it.

These blunders were immaterial in the short-term because Gould's publication of the Australian rat achieved nothing beyond increasing the size of his vast book. Far from becoming the basis on which new discoveries about the rat were organised, Gould's text and plate were ignored. During the next forty years, just two scientists at the Australian Museum referred to *Mus longipilis* and neither added to what Gould had reported. Otherwise, the many colonists who wrote about the rat did so without reference to Gould, since his *Mammals* primarily found subscribers in Europe and the twelve copies within Australia were confined to scientific societies, public institutions and the very wealthy. Gould's *Mammals* brought the rat to an international audience for whom it meant little or nothing, but failed to reach settlers trying to make sense of it.

CHAPTER 3

A Wonder of Art

Colonists named two places after the long-haired rat, both out of horror. The first was on Cooper Creek, 120 miles west of where Thomas Wall collected his specimen in 1847. When the Victorian Exploring Expedition led by Robert O'Hara Burke and William John Wills reached the Cooper in November 1860, they wanted to establish a depot where they could leave some of their men, food and equipment while they sought to cross the continent. A big waterhole appeared ideal by day, lined with trees, surrounded by grass, and rich in waterbirds and fish. But, at night, the explorers found themselves under siege. Wills recorded that they were 'forced to remove' by rats that attacked their stores in such numbers that the men 'could keep nothing from them, unless by suspending it in the trees'. Burke reported that they were 'driven out by the rats'. Although they often identified this site as Camp 63, meaning their sixty-third since leaving Melbourne, the explorers also called it the 'Rat Hole'.

The party immediately moved downstream, hoping the rats were confined to Camp 63. Another waterhole, which they reached two days later and became the explorers' Camp 65, looked promising. Wills considered the surrounding sand ridges relatively 'agreeable' because of their 'freedom from flies, ants, mosquitoes, and rats', but at night the waterhole, too, was thick with rats. Burke recognised their abundance would make Camp 65 'a very disagreeable summer residence' but, after the explorers' experiences at the Rat Hole, he decided it was as good as they could hope for.

The Burke and Wills expedition left many records of the rat—writing about it in journals and reports, painting it and testifying about it before a commission of inquiry. The explorers typically did so because they thought of the rat as having blighted the expedition, which saw four of them become first to cross the continent but seven die. When one of the survivors revealed that the rat had erased traces of Burke and Wills after they returned from the Gulf of Carpentaria—so other members of the expedition did not know their leaders had revisited Camp 65—there was immediate evidence that the rat was one cause of the explorers' deaths. But colonists ignored it because they were preoccupied with identifying a human cause for the deaths of Burke and Wills. Colonists also failed to ask why the explorers almost never ate the rat. The answer was too obvious. 'Civilised' people did not eat rat. Yet had the men eaten it when starving, the long-haired rat could have been their salvation.

Burke and Wills first sighted the rat after they left country occupied by Europeans and made their way, with six other members of the expedition, from the Darling River to the Cooper, through land that appeared 'a splendid grazing country' to Burke in the wake of an exceptionally wet winter in the interior. There was just one night when the party found no water. Otherwise there were plenty of shallow pools, though some were drying fast. As they came on the watercourse that the local Wanggumara people called the Bulloo, they saw everywhere 'indications of the most violent floods'. A few days later, they passed a dozen Aborigines catching snakes, iguanas, mice and rats.

The explorers soon found the rat closer to the Darling, near the lake that would become known as Torowoto and also north of the Cooper, where the sand ridges were 'so honeycombed… by the burrows of rats, that the camels were in danger of falling'. Over the following months, the men encountered the rat both in well-watered country and in terrain so dry the men had to carry water for miles. The rat's survival amidst this aridity aroused the curiosity of Hermann Beckler, one of the party's German officers. He surmised, correctly, that the rats had 'water sources, known perhaps only to them' or knew 'how to replace water with something else'.

The encounters between the explorers and the rat depended on the speed at which the men travelled and the quantity of food and equipment they carried. If they used a campsite for just one night or had few supplies, they saw little of the rat. If they stayed longer at a camp and were heavily laden, the rat made its 'presence obvious' on the second night, causing 'great annoyance' and soon appeared in greater numbers. It was characteristically curious, quick to investigate strange objects, attracted by the men's flour, sugar, dried meat and anything made of leather. As the explorers first experienced on the Cooper at their Camps 63 and 65, they came under sustained assault from dusk to dawn.

These assaults were novel. Edmund Kennedy and his men did not experience them on the Cooper, as they had temporarily buried most of their flour and sugar so their horses' loads would be as light as possible, and they struck the rats within a small area for just a few days. Because Aboriginal dwellings contained much less food and equipment than the explorers' camps, they too were little of a lure for the rats. Even the settlements of the Yandruwandha people on the Cooper, which included dozens of dwellings and were semi-permanent, would have attracted relatively few. While a vast array of native animals were passive victims of the European invasion of the continent, the long-haired rats fought back. According to William Wright, who joined the Burke and Wills expedition on the Darling to lead one of its many subdivisions into small units, the rats were 'far bolder than the majority of domesticated animals in their attacks'. Possibly no other Australian animal—not even the dingo—responded so aggressively to Europeans.

The rats jumped and climbed over the men as they tried to sleep, bit them and fought 'bloody duels' among themselves, sometimes on the explorers' blankets as the men lay beneath. The rats also attacked the men's supplies and equipment, eating through sacks of flour and bags of dried meat, gamboling about in biscuit barrels and chewing through saddles, stirrups and shoes with the 'wildest destructive rage'. The rats negated one of the men's rare successful attempts to kill birds, stealing their catch. They also frustrated the explorers' one attempt to grow vegetables, eating the seedlings as they appeared. The men only succeeded in stymieing the rats when they suspended their food and equipment from trees.

The animals that feasted on the rats included birds, identified by Hermann Beckler and William Wright as hawks. An array of snakes ate the rats too. After setting out from the Darling for the Cooper in January 1861, Wright's party encountered many western brown snakes, one of Australia's deadliest species. They also saw—and killed—a snake that scientists have identified as Stimson's python which, like all pythons, relies on constriction to kill its prey, but the explorers thought 'most poisonous'.

The rat also presented an opportunity for the explorers because it was the one source of meat that regularly came to them. Whether the men exploited this opportunity was largely determined by the Irishmen, Englishmen and Germans who dominated the expedition. But its two Afghans, Dost Mahomet and Belooch, who cared for the party's camels on their journey from Karachi to Melbourne, then were hired to continue looking after the camels during the expedition, also had to decide what they would eat. So did John Smith, a former station hand on the Darling, whose mother was Aboriginal and father was British.

Rats were eaten only by the very poor in mid-nineteenth century England. Their prime use was in 'sporting public-houses' where a terrier would be put in an enclosure and there would be bets taken on how many rats the dog could kill in a set period or which of two dogs might kill more. As these 'rat pits' attracted big audiences of men, and some women, and the most successful dogs became renowned for their killing, a trade developed in supplying the rats, which were usually caught in the countryside and brought into towns and cities in wire cages. In his celebrated book *London Labour and the London Poor*, Henry Mayhew calculated that more than 100,000 rats were killed in these pits each year in the capital.

James Rodwell, the principal British writer about rats in the 1850s, fixed on how they caused great economic loss. He called for 'universal warfare' against rats to curtail their ravaging of 'barns, granaries, ricks, mills, cornfields, warehouses'. Yet far from suggesting that rats carried disease, which Europeans barely understood, Rodwell maintained they contributed to public health, by devouring 'animal matter thrown into drains'. He even claimed that, without this scavenging, there would be 'a constant plague in London'. He also recognised that rats were not just edible but 'cheap and nutritious', offering the 'most delicate and sweet eating'.

Colonists went in big numbers to Australia's first rat pit, established in Melbourne in 1859 in inner-city Collingwood, where 'sports' nights were staged every Wednesday and Saturday. They ate rats only out of desperation, though there were other animals that they were even more loathe to consume. The explorer Edward John Eyre was an example as he headed west along the Great Australian Bight in 1841 with an Aboriginal man whom he did not bother to name and described demeaningly as a 'native boy'. For several days, this Aboriginal man and Eyre were on very limited rations with their provisions near exhausted. Then they killed a bandicoot, a rat and a snake—more food than they required. They ate the bandicoot and the rat, but left the snake.

When Burke and Wills set out from Cooper Creek in December 1860, they left three men at the depot at Camp 65 under the command of William Brahe, another of the expedition's Germans. As they relied on the supplies carried from Melbourne, their usual diet was a pint of boiled rice with lots of sugar for breakfast, damper with a little salt pork or beef for lunch and then two biscuits and a pint of tea each evening. They also caught ducks that congregated on the waterholes closest to the depot, until the birds became 'shy' after 'constant firing', and the men 'got careless about them'. The rat was an obvious alternative if the men wanted fresh meat. But their killing of about forty each night was a vain attempt to reduce the rat's numbers and a form of revenge for the misery it caused, not a mode of extending and enriching their supplies.

Brahe's party quit Cooper Creek in April 1861 either a day or two before Burke and Wills returned from the Gulf of Carpentaria, after crossing the continent. Brahe left Camp 65 partly because his men were succumbing to scurvy. He also wanted to leave a substantial cache of food for Burke and Wills—indicating where it was buried on the soon-to-be famous 'Dig Tree'—while taking sufficient supplies to sustain his party until it reached European settlement on the Darling. Had Brahe and his companions made the rat a regular part of their diet, they would not have had to leave for this reason. When Burke and Wills returned to the depot, Brahe's party would still have been there. Burke and Wills would have survived, if not in legend.

The party led by William Wright got closer to eating the rat after setting out from the Darling in January 1861. Hermann Beckler, the expedition's doctor, was the catalyst, as he embraced the common belief that any fresh food—including meat—could stop scurvy. When his companions began sickening, Beckler viewed the rat as a cure. He argued that its flesh would be 'as delicate, tender and fresh as could be found anywhere', presenting no risk of disease because of 'the cleanly life of these creatures, their harmless, unspoilt food and their abode in the unsullied lap of mother nature'. When the men reached the Bulloo River, where Wright killed up to fifty rats a night using a trap he made, Beckler proposed they eat some. While all his companions protested, Beckler dismissed their objections as prejudice and prepared a dozen 'in the European manner', skinned and gutted.

These long-haired rats were in a large tin dish, looking 'just like a row of little piglets' to Beckler's delight, when Wanggumara people entered the explorers' camp, indicated that the land belonged to them and told the explorers to leave. As they stayed put, Wright may have offered some of the rats to the Wanggumara's leader in an attempt to mollify him. Alternatively, the Aborigines' leader may have taken the rats from the dish, then dropped them as the confrontation intensified, only to pick up one, make 'a harangue upon it', and conclude 'by flinging it contemptuously' at the invaders. Ultimately, according to Beckler, 'The twelve rats which I had prepared so prettily...lay on an anthill, and our first meal of rat was to be no more.'

The expedition's saddler, Alexander Macpherson, and Trooper Lyons of the Victorian Police were the first Europeans to eat the long-haired rat, having run out of food and water while trying to bring despatches to Burke. They were so desperate that they ate the flesh of a long-dead horse, which caused Macpherson to faint, and once 'rinsed out their mouths with their own urine, and derived great relief from it'. After that, the sight of numerous rats at Torowoto was like manna from heaven. After 'a good deal of difficulty...sitting up all night', Macpherson caught three, which Lyons and he ate immediately.

Burke and Wills were also desperate after they returned in April 1861 from the Gulf of Carpentaria to Cooper Creek with their one surviving companion John King, and found that Brahe's party had just left. Exhausted and short of food, they turned to the local Yandruwandha people who fed them fish and bread made from the seed of nardoo, the native clover fern. In May, when Burke, Wills and King were on the Cooper downstream of Camp 65, the Yandruwandha offered Wills 'a couple of nice fat rats' baked in their skins, with their tails already pulled off, which he was surprised to find 'most delicious'. Wills recognised that Burke, King and he needed to devise a means of trapping the rats, but they failed to do so. While probably stymied by their limited bush skills, their scant supplies and equipment reduced their chances, since the rats were not attracted by poverty.

The rats were still at Camp 65 in big numbers when William Wright and William Brahe visited it in May, without realising that Burke, Wills and King had been there. Wright and Brahe remained oblivious because Burke, Wills and King had tried to leave the depot as they found it and had not recorded their return on the surrounding trees. The rats had also erased the footprints of Burke, Wills and King and any other marks that they left around the depot. When a commission of inquiry quizzed Brahe about what he saw there, this exchange occurred: 'Did you see any impression of human feet?' 'No impression.' 'Why?' 'From the number of rats.'

The rats had probably declined by September when a search party reached the Cooper. Its members made no mention of extraordinary numbers of rodents. But the party's deputy leader, Edwin Welch, who was like many colonists in characterising an array of small Australian animals as 'rats', recalled that rats were fat, healthy and plentiful on the creek. When the Yandruwandha offered the search party some to eat, its members immediately accepted, though they had ample stores of their own. They did so, in marked contrast to members of the original expedition, because their leader, Alfred Howitt, 'would sample anything in the interests of science'.

Wills recorded that the Yandruwandha used a 'large boomerang'—most likely, a sword club—'for killing rats'. Welch, who was much more interested in Aboriginal culture, recalled that rat catching was a preserve of just a few Yandruwandha—not only women, who typically pursued small animals, but also men, who generally pursued bigger game. The most striking aspect of Welch's account is that it did not involve the Yandruwandha digging out the rats. Instead, they used song to enchant the rats and get them to leave their burrows. Welch recorded: 'They merely seated themselves close to the burrows, armed with a waddy, or light club' and sang 'in low, weird and dirge-like tones' that 'the rats could not withstand and, as they came from their holes, they were promptly knocked over and cast aside until a sufficiency had been obtained'.

Ludwig Becker, the expedition's artist, zoological collector and second German officer, was the one member of the original party who had probably seen the rat before the men set out. He would have done so in either Melbourne's Public Library or its National Museum, which both subscribed to John Gould's *Mammals of Australia*. Because Becker was an artist-scientist, he probably examined this lithograph when it arrived in Melbourne in 1856. As he prepared to go on the expedition in 1860, he may have looked again to heighten his familiarity with species Europeans had already classified. But there is nothing to suggest Becker remembered Gould's plate when he headed beyond the Darling. More likely, he thought the rat new to Europeans.

Had the expedition been properly fitted out as a scientific venture, in keeping with its organisation by the Royal Society of Victoria at prodigious expense, Becker would have been carrying equipment for catching and preserving animals such as the long-haired rat. But in the confusion and chaos that engulfed the expedition, the Society did not provide Becker with any when he set out. Although he immediately pointed out this failing and requested traps, preserving bottles, alcohol and arsenical soap, the Society still did not provide them. Almost five months later, as Becker headed inland from the Darling River in January 1861, he lamented that he had 'only an outfit consisting of a few colours and sketch-books and two geological hammers'. With no means of preserving a physical specimen of the rat, Becker's only option, when he encountered it in February, was to paint it.

There were no trees or bushes to provide shade. The heat was intense, reaching well over 100 degrees Fahrenheit (38°C). There was nothing to drink, despite first appearances. According to Hermann Beckler, a marvellous island often shimmered before the men, rising out of a peaceful, blue sea. The explorers had 'the magic of the most wonderful water on all sides'. But it was 'an illusion' as behind lay 'nothing but sterile, red hills sprinkled with white limestone'. The men could stay there only by fetching water from one of their old campsites more than fifty miles away. At worst they had nothing to drink, at best a spoonful an hour during the day, followed by half a cup at night. Hermann Beckler called this spot 'Dry Camp'.

The flies, which appeared after sunrise and disappeared after sunset, made this campsite worse. They pursued all the men, besieging their plates, drowning in their tea and slipping into their mouths with their food, as well as getting trapped in the corners of the men's eyes in their search for moisture. Yet they gave Ludwig Becker most cause to despair by interfering directly with his painting. They 'sucked the colours and inks from his quills and brushes and threw themselves recklessly on to every damp spot on his paintings'. Becker's name for this site was 'Desolation Camp'. When the temperature there reached 112 degrees Fahrenheit, he described this camp as 'a very hell'.

His one view of the surrounding landscape, looking across the plains at noon, included two dingoes. While these dogs had been in Australia for just a few thousand years, Becker's contemporaries thought they were either an indigenous species or one introduced by the first Aboriginal people. The dingoes painted by Becker would have been attracted by the rats, which emerged from their burrows after sunset, giving their characteristic high-pitched whistle, which Hermann Beckler considered 'the wildest and most unhowely noise'. As the rat set upon the explorers and laid waste to their food and equipment, Beckler thought of it as 'the undisputable owner of those dreary grounds'. William Wright concurred when he called this campsite 'Rat Point'—the second, and seemingly last colonial place named after the long-haired rat.

The contrast with H.C. Richter's situation when he drew his plate for Gould could not have been greater—and not just because Richter worked in comfort in London. Richter had also seen only one rat and this specimen had been dead six years. He had no idea what the country inhabited by the rat looked like and had never heard about, let alone experienced its depredations. But Becker had seen hundreds if not thousands of them, male and female, adult and juvenile. He also knew much more about them than he could convey in a single watercolour, particularly a tiny one, and had to decide whether to include anything of the rat's environment and of his own experience.

Becker was a master of nocturnal subjects, easily capable of depicting how the rats overran the explorers' camp at night and attacked the men. But as with most of his work before and during the expedition, he opted to follow the conventions of natural history illustration and divorce the rat from its context. Much as Richter had shown the rat from the side and frontally, Becker depicted it in profile together with a frontal view of its head and two details of its claws. His text was primarily an elaboration of the rat's colouring, which he described as generally 'dark ash-grey'. Unusually, he did not record his specimen's size or gender. He gave little sense of its habitat, simply describing it as 'on border of the mud plains'. He wrote nothing about its numbers and assaults.

Since one of the goals of the expedition was to enhance knowledge, Becker's watercolour may be seen as a failure. He conveyed little of what he knew. Yet the painting may also be seen as a triumph because of the way Becker transcended the personal. He gave no hint that the rat had been despoiling the camp, let alone that he himself had experienced its assaults. If anything, the painting is sympathetic. That the fifty-one-year-old Becker was able to render the rat so beautifully, in such bleak circumstances, when he is said to have ruined his health through excessive drinking and was beginning to suffer from the scurvy that killed him two months later, makes it one of the wonders of Australian art.

Because of Becker's exactitude—a feature of his work—the painting clearly identifies the rat as the species encountered by Kennedy's expedition, collected by Wall and classified by Gould as *Mus longipilis*. Had Becker's watercolour been published in an illustrated account of the expedition, as he expected, that would have been obvious. Colonists would have known the long-haired rat could sometimes be found in vast numbers deep in the interior. Had they examined the explorers' journals, they might have associated this irruption with the aftermath of big rains and floods. But because neither Burke nor Wills lived to write a book about the expedition, Becker's watercolour remained unpublished along with all his other paintings, and no one realised for more than a century that Burke and Wills had encountered the long-haired rat.

CHAPTER 4

Bush Naturalists

Within a few years, there were more rat plagues not only in the Channel Country but also in the Gulf Country of Queensland, the northern interior of South Australia and the central west of New South Wales. These irruptions were the stuff of just a few brief, scattered reports, soon forgotten. But the rat was remembered by those who had experienced it and, when offered the opportunity to write about these episodes, some colonists did so in compelling detail. Their forum was the weekly *Queenslander*, published in Brisbane, which, in 1878, established the first regular newspaper column devoted to Australian natural history.

Its mainstay, Price Fletcher, styled himself the 'Bush Naturalist' because he had no formal scientific training. Although he wrote for the *Queenslander* from near Brisbane, most of his articles were about the colony's north-west interior where he had managed a pastoral station. His great passion was birds—not just observing and describing them but trying to stop their demise. He also identified species for his readers—sometimes responding to their descriptions, sometimes their drawings and sometimes specimens, which included the ovaries and foetuses of assorted marsupials. Twenty years after the publication of Charles Darwin's *On the Origins of Species*, when its arguments about evolution were still widely rejected in Australia, Fletcher revealed himself as a Darwinian who believed in 'the origin of variety through the power of natural and sexual selection'. He was alive to the colonists' immediate impact on the environment and deplored their excessive exploitation of it. 'Never after the country is once stocked is it again seen to perfection,' he observed. 'Our greed to get rich is too great.'

Natural history was so popular in Britain in the mid-nineteenth century that it has been dubbed a 'national obsession'. It lacked the same appeal in colonial Australia, but Fletcher's 'thrilling' pieces, as one of his admirers characterised them, found an appreciative audience. His was '*the* article' for which some readers 'thirsted as mail-day drew nigh'. His 'graphic writing' made the *Queenslander* their 'first favourite'. He also vetted contributions from readers, who ranged from admitting almost complete ignorance of natural history to boasting about their expertise, having spent years in the bush scrutinising 'anything strange in the habits of either birds or animals'. Fletcher encouraged, instructed and admonished this diverse group of men—as they all seem to have been—and invited them to address particular subjects, resulting sometimes in months of correspondence. Until 1881, when he moved to Mackay in north Queensland to manage a sugar plantation, Fletcher was the unrecognised environmental counterpart of Australia's renowned literary ringmaster of the late nineteenth century, J.F. Archibald of the Sydney *Bulletin*.

Fletcher's own editor for most of this period was Carl Feilberg, a Danish journalist, who fixed unprecedented attention on the colonists' attacks on Aboriginal people before losing his post due to his outspokenness. At the start of a series of articles in the *Queenslander*, soon republished as a pamphlet, which Feilberg titled ironically *The Way We Civilise*, he wrote: 'On occupying new territory the aboriginal inhabitants are treated in exactly the same way as the wild beasts or birds the settlers may find there. Their lives and their property, the nets, canoes, and weapons which represent as much labor to them as the stock and buildings of the white settler, are held by the Europeans as being at their absolute disposal. Their goods are taken, their children forcibly stolen, their women carried away, entirely at the caprice of the white men.'

In December 1878, Fletcher wrote of 'a species of field rat' that, 'creditable witnesses' assured him, had become 'a perfect Egyptian plague' a few years before over Queensland's north-western plains. Fletcher had heard that the rats 'really seemed to permeate from the earth, for every crack in the ground contained a nest in them; they penetrated the camps, tents, and houses, up into the drays even; they ate saddles, hobbles, boots, damper, everything and anything that was not hung up'. The rats abounded for so long that 'cats got sick of them; dogs, which when they first appeared, eagerly worried them, took a disgust, and would not stir even if one came within a yard of them'. Then, just as the rats appeared suddenly, so they vanished, 'how, or why, or where...no one knows'—a pattern identified by Fletcher as 'characteristic of Australian natural history'. Far from thinking himself lucky not to have encountered the rats, Fletcher 'often regretted' that he 'was not an eye-witness' and called on 'some old resident to write an account of this visitation'.

The responses recording events longest ago were of plagues in 1863 and 1864, recorded only sketchily at the time. One report was of a landscape of abundance in the Gulf of Carpentaria where rats, snakes and iguanas were 'so plentiful' that local Aborigines grew exceptionally fat. A second described a cattle station near Lake Blanche in the South Australian interior as 'completely overrun with rats'. A third, from a party travelling from the Darling River to the Gulf Country and back, was that the rats encountered by the Burke and Wills expedition were 'more or less numerous the whole way across the continent'. In a letter to the *Queenslander* in 1879, a bushman who had been on the Flinders River near the Gulf recalled that the rats were a 'scourge' in 1863, 'a wet year, with its flood in June', and 'a perfect nuisance' in May 1864. Kenric Harold Bennett wrote at greatest length about the plague in 1864 and, unusually for Price Fletcher's contributors, did so under his own name rather than a nom de plume. This choice may reveal more than that Bennett was happy to disclose his identity. Perhaps unconsciously, perhaps consciously, Bennett wanted to be known for his understanding of Australia's natural history.

Born in Hobart in 1835, but raised in Gippsland, Victoria, where his father had a cattle station, Bennett revealed his eagerness to encounter what was new to Europeans when he applied unsuccessfully to accompany Burke and Wills in 1860. Bennett soon got an opportunity to work near the colonial frontier through his uncle William Adams Brodribb, a pastoralist who came to specialise in securing land available from government at negligible cost because it appeared waterless. In 1861, Brodribb tendered successfully for five blocks inland from the Lachlan River in central western New South Wales. Because colonists had never occupied this land, Brodribb was directly involved in dispossessing its Wiradjuri owners even as he embraced a Wiradjuri word in calling these blocks Moolbong, said to mean 'grass gone away'. His prime agent was Bennett who inspected this land for him, then, with his father, sank a well that struck good water at a depth of more than 100 feet.

In 1863 Bennett's parents acquired 35,000 adjoining acres, which they called Yandembah, a Wiradjuri place name, even as they too were displacing the Wiradjuri. As water was again an issue, the Bennetts sank a well to supplement the one small, intermittent creek and built a house on a low sand ridge out of the local callitris pine, which they otherwise preserved because trees were scarce. Anyone who 'wilfully destroyed' one of these pines 'would be regarded as a goth', Bennett observed of colonists there. A journalist-cum-drover, who wrote a series of articles about this region in 1877, judged Yandembah the 'prettiest place' in the 'back blocks'. The women of the Bennett family were integral to this traveller's astonishment and delight at 'the sudden transition from a life of savage barbarity, and brutalised natures, to an abode of refinement, peace and virtue', prompting him to stop for two days of 'music, good dinners, gossip and croquet'.

Bennett, meanwhile, managed Moolbong which Brodribb stocked with 23,000 sheep. When almost no rain fell for more than a year, half the sheep and all their lambs died. But some years were bountiful, vindicating Brodribb's speculation. The best were 1864 and then 1870, when Moolbong received sixteen inches of rain between January and June, enabling Brodribb's flock to swell to 50,000. Emboldened by this success, Brodribb set Bennett searching for more land and, in 1872, secured three more waterless blocks—even further back from the Lachlan River—which had been abandoned by other pastoralists. Brodribb called these blocks Moolah, another Aboriginal word, as well as a common spelling of 'mullah', widely used in texts about Islam. Bennett also travelled much further for Brodribb and, on an expedition to Queensland's Darling Downs, met Hannah Dunne, usually known as Annie, who was thirteen years his junior. They married in 1875 when Bennett was forty, but they were rarely together. Because of Yandembah's relative comfort, Annie generally lived there, with their children.

Bennett managed Moolbong until Brodribb sold it at a big profit in 1875 when stations like it and Yandembah were fetching almost as much as those with river frontage. Then Bennett became manager of Moolah, a much tougher assignment because of its enduring dearth of water. But the journalist-drover, who had delighted in the refinement of Yandembah in 1877, was impressed by what Bennett was creating on Moolah with its new homestead 'neat and compact, with outbuildings and yards to match'. There was also 'more life and stir' than on other properties: 'carpenters, bush hands, tank-sinkers, boundary riders' with 'horses, sheep, bullocks, sawing, neighing, bleating, and yabbering on all sides'. Bennett remained there until Brodribb sold Moolah, at a loss, late in 1879, and Bennett moved to Yandembah, which his younger brother, Edward, was running following their father's recent death.

Bennett described himself as 'a great lover of natural history', 'very fond of...all its branches'. He recognised collecting as not just his 'favourite pastime' but a 'sort of mania'. Already as a boy in Gippsland, he roamed the scrub near the foot of Mount Baw-Baw, shooting birds for their skins, collecting their nests and eggs and keeping a sharp eye on much else, including tree snakes and tree frogs. At home, on his family's cattle station, he raised a kangaroo that followed him around 'like a dog' until a dingo got into its shed and killed it. When he arrived back of the Lachlan, Bennett delighted in the many species new to him, perhaps above all the flock bronzewing, a native pigeon. He would recall how 'countless multitudes' of bronzewings were found on the plains with 'immense assemblages' attracted by the few water tanks and dams sunk by the settlers. Echoing accounts of the passenger pigeon in North America, he recorded how, when alarmed, 'the whole flock would simultaneously take wing, the noise resembling thunder, and as the immense cloud wheeled around preparing to again alight, they would fairly darken the air'. Because kookaburras were among his 'great favourites', and they were not found in the back country, he took a couple of young from the river and raised them in 'a sort of semi-domestication'. He reared mallee hens, which many colonists considered impossible, then wondered why he had bothered when they displayed 'no intelligence or affection'. He established an aviary, filled with the most colourful native birds, only for his kookaburras to kill many, prompting him to release them.

Bennett learned much from the Wiradjuri. Otherwise, he was an isolate. Unable to afford books and without access to a library, Bennett could not evaluate the significance for European science of what he saw and found. He was in no position to identify anything as the first or the only. While he knew which species were rare back of the Lachlan, he often had no idea whether they were rare or common elsewhere. His knowledge of the names of birds made him even more of a colonial oddity. He knew what many birds were called in two, perhaps three, Aboriginal languages. But he often did not know their English names since there were many species that, as far as he could determine, had been given no names by colonists or 'if they have no one about here knows them'. He also recognised that, insofar as settlers in the back country had named these birds, these names were not worth adopting as they were not in wider colonial currency and hence would not communicate anything. While aware of the Linnean system of classification, Bennett knew the Latin names for only a few species.

His was a private passion until Fletcher's column debuted in the *Queenslander* in 1878 and occasional copies reached the forty-three-year-old Bennett at his 'out-of-the-way corner of the country' at Moolah. When he first submitted a piece about birds to the *Queenslander*, at the start of 1879, Bennett hoped not just to see himself in print but also to become 'better acquainted' with Fletcher whom he recognised as a 'kindred spirit'. While Fletcher sometimes dismissed would-be contributors for failing to provide 'new facts', he expressed his 'best thanks' for Bennett's 'interesting communication' and encouraged him to write more. Soon Bennett was Fletcher's only regular contributor from New South Wales, while the rest were from Queensland. Just as Bennett would contradict, qualify or expand on their pieces or ask them for information, so would they of him. While one who dubbed himself 'Thickthorn' would reveal himself a few years later as Charles Walter de Vis, a Cambridge graduate who had studied zoology and been the curator of a small museum in Manchester, the others most likely were all bush naturalists. To Bennett's manifest delight, he found himself part of an informal scientific society, its members all passionate about Australia's environment, eager to communicate to a general audience.

Bennett often had too much to say. One of his letters about snakes and birds, published in June 1879, was so long that Fletcher was 'compelled to condense it very considerably', though its published version was still more than 800 words. Bennett's letter about rats, which was about the back country of the Lachlan, not Queensland as Fletcher had anticipated, ran to more than 1500 words as Fletcher published it that July. Fletcher acknowledged it as an 'interesting contribution' but again was 'obliged to suppress a portion'. To avoid a recurrence, Fletcher advised: 'Should the writer again favour us, we must beg of him to condense his communication as much as possible.' Otherwise, Fletcher's one instruction was: 'Please write only on one side of your paper.'

Bennett recalled that, for between six and eight months in 1864, rats were found in 'incredible numbers all over the vast plains' far from the Lachlan, which stretched sixty miles wide that year when it flooded. While some of the rats burrowed into the soft soil, others made nests of grass in dwarf saltbushes that covered the back country. As usual, they attacked the colonists' food and equipment, especially their saddles. In response, the settlers attacked them in new ways. They spread poison—most likely strychnine—and introduced cats in an attempt to control them. Birds of prey and dingoes also feasted on the rats. Bennett found eleven in the stomach of one dingo—a prodigious figure. But there was 'no perceptible diminution' for months. Then, 'all at once' the rats 'greatly decreased, and within a week...not a rat...was to be seen'.

Bennett was puzzled by wooden structures, which were usually on scrubby sandhill and pine ridges. Up to five feet in diameter and three feet high, these structures were made of sticks so firmly woven together they 'could be turned over without falling to pieces'. Bennett initially thought they contained the nests of the rats that were in plague proportions, but when he investigated, he found that none was occupied. Nor was there anything that indicated their 'occupation by any animal past or present, whilst the interstices between the sticks were so small that it would be scarcely possible for a mouse, much less a rat, to pass through'. When he asked the Wiradjuri, they explained these structures were made by a species something like a small kangaroo rat, once numerous, which had disappeared.

Another reader of the *Queenslander*, who had also been back of the Lachlan, responded that these structures were 'built by rats'. Having found their nests within the woven sticks, he doubted whether Bennett had investigated them properly. While he remembered their makers being 'about the size of an ordinary bush rat', he knew nothing further, explaining, 'I am not a good enough naturalist to say what sort they are.'

These nests, much like those of badgers or pack rats in North America, were a protection against predators and a means of temperature control. The earliest European account of them came from Thomas Mitchell who saw them near the confluence of the Darling and Murray rivers in 1835. As Mitchell described them, they were 'formidable fabrics' worked 'around and interlaced with some small bush', with an occupied nest at the heart of each one he examined. Charles Sturt encountered many more in the same area in 1844, then went on to see 'hundreds of these nests...generally inhabited by from six to ten rats and...not unfrequently five feet high'. While John Gould initially identified this species as a *Mus*, he reclassified it as a *Hapalotis*, the name scientists applied to an array of Australian rats until they discovered that the German entomologist Jacob Hübner had already applied it in 1821 to a genus of moths.

The best early account of these rats came from Gerard Krefft, one of the many Germans who loomed large in Australian art, science and exploration in the mid-nineteenth century. Twenty years after Thomas Mitchell recorded the stick nests at the junction of the Murray and Darling rivers, and a decade since the occupation of this land by pastoralists, Krefft was there in 1856–57 as a member of a Victorian Government party led by another German, William Blandowski, who was the colony's zoologist. Just as this position was a novelty in Australia, so was the expedition: the first government party sent out purely to examine a region's fauna. Krefft, who was second-in-command, seized this opportunity to study the impact of European settlement, with Mitchell's collecting providing a yardstick. Krefft was in an exceptional position to do so since he did not just pass through the area but stayed eight months and worked with a team of 'devoted natives' from the local Nyeri Nyeri people, who became his 'permanent huntsmen' in return for payment. As Krefft described it, 'the boys would go to collect insects, the women to look for smaller mammals, and the men looking for larger game'.

Krefft began interpreting the results of this expedition after moving to Sydney to become assistant curator, then curator—effectively, director—of the Australian Museum, which had opened to the public in 1857. A paper about his work at the confluence of the Murray and the Darling, delivered to the Philosophical Society of New South Wales in 1862, has become the foundation of an array of studies exploring the causes of the decline and extinction of species unique to Australia. Krefft's work, with its unusual empirical richness, has underpinned the idea that 'the impact of Europeans on the Australian environment is not measured in centuries, but in decades' with many species disappearing locally within thirty, twenty or even ten years of an area being settled.

Among the species Krefft fixed on were the greater and lesser stick-nest rats, then known to science as *Hapalotis conditor* and *Hapalotis apicalis*, now known as *Leporillus conditor* and *Leporillus apicalis*. They both had compact bodies, large ears and blunt noses, but the smaller of the two had white hairs, especially on its tail tip. Their nests were typically located in watershed and drainage areas that supported perennial, semi-succulent or succulent shrubs, including salt-bushes, blue bushes and pigfaces. While Mitchell had found the greater species south of the Murray, Krefft could not and its only traces were 'a few empty nests'. To the river's north it was rare. Krefft believed it had 'already retreated before the herds of sheep and cattle'. He predicted it would 'soon be extinct' which was true of that area and ultimately of the entire continent, though not the Franklin Islands off the South Australian coast where this greater species would survive, allowing its eventual reintroduction to the mainland. Krefft saw no immediate threat to the smaller species, which had colonised some of the nests of the bigger, but it became extinct in the mid-twentieth century. Because of his particular interest in these rats, Krefft kept some 'more or less as pets and often fed them on his damper' though, as part of experimenting with eating native species, he also ate one and considered it 'excellent'.

Krefft, who had aspired to the position on the Burke and Wills expedition secured by Ludwig Becker, was also the one nineteenth-century scientist to wonder about the identity of the rats encountered by the Victorian explorers. While nothing like as accomplished an artist as Becker, Krefft was a much more substantial scientist, but he erred when it came to identifying the rats found between the Darling and the Cooper and beyond in 1860–1861. Having read the explorers' published accounts but not seen Becker's drawing, Krefft had 'no doubt' the animals encountered by the Victorian expedition at Rat Point were lesser stick-nest rats.

Bennett was oblivious to Krefft's work. At Moolah and Yandembah, he had no access to the *Proceedings of the Philosophical Society of New South Wales* or Krefft's catalogue of mammals in the collection of the Australian Museum. But if Bennett's investigation of the stick nests was as careful as his account suggests, they were empty in 1864. Within just two years of colonists occupying the back country of the Lachlan, the stick-nest rats were gone—most likely, due to the colonists' livestock, as Krefft suggested for the area at the junction of the Murray and Darling while offering an even more dramatic instance.

The animals that abounded in 1864 were long-haired rats, which had multiplied and spread in the wake of the prodigious rains. One clue to the rats' identity is their appetite for the colonists' saddles, a favourite of the long-haired rat. Another clue is the identity of the birds that primarily preyed on the rats. Although Bennett would not realise it until the end of 1879, these birds were letter-winged kites, the species most closely associated with the long-haired rat. Without Price Fletcher asking for recollections of rat plagues—and without Kenric Harold Bennett writing from country that Fletcher never envisaged—there would be no extended account of what occurred.

CHAPTER 5

A Perfect Egyptian Plague

Rats excited exceptional attention in 1870 due to the Siege of Paris. As the Prussian army of King William I surrounded the French capital, seeking to starve its residents into surrender, any form of meat became a rarity that only the well-to-do could afford. Kangaroo from the city's Jardin d'Acclimatation was one novelty. Rat was another. While marginal to the Parisians' diet, its consumption became a symbol of the extremity of their situation. There had never been so many accounts of rat catching, studies of the rat market or choices of rat recipes. Some dishes were designed to mask the rats' taste through sauces and condiments, others to reveal it, which was said to be like 'a cross between pork and partridge', if not 'something between frog and rabbit', with 'the delicacy of a tender chicken'. Either way, rats were judged 'very good', even 'excellent' to eat, though their consumption was still deemed 'barbarous'.

Rat jokes, rat poems, rat stories and rat art proliferated too. An early quip was: 'What is a rat?' Answer: 'A future partridge.' In Théodore de Banville's 'Ode to the Rats', the animals blamed their killing on the Germans laying siege to the city and declared they would find revenge by invading Prussia and destroying all they could. *Le Figaro* reported that 'a man was puzzled to find himself pursued through Paris by a pack of barking dogs, until he remembered that he had eaten rat for breakfast'. The acclaimed illustrator Cham showed a queue of Parisians, kneeling in a gutter, waiting their turn to catch any rats that might emerge from a drain. At the Paris Salon in 1872, Narcisse Chaillou exhibited a painting of a vendor rolling up his sleeves to skin a rat on sale for two francs.

Rats were irrupting in Queensland and then South Australia while Paris was besieged. Rats were also said to be in immense numbers in New South Wales, though the plague there was probably limited to the introduced mouse, *Mus musculus*, which was spreading inland with increasing impact on crops. Sydney's *Town and Country Journal* linked the events in the French capital and the Australian outback when it observed of the local rats: 'They are in everything and everywhere and in sufficient quantities (had they been in Paris) to have enabled the French to hold on for another month. There they would have been a god-send, but here they are a pest and a nuisance.'

Price Fletcher was thinking of the irruption that peaked in Queensland in 1870 when he asked his readers to submit accounts of the 'great Egyptian plague'. It had gone unrecorded when it occurred but, as Fletcher expected in 1878, there were those who remembered it. Now they had an invitation to write, and they seized it, providing the first detailed chronology of such an episode, corroborating each other in many respects but also arguing about some of what occurred. After publishing their letters, Fletcher observed with justified self-congratulation: 'It was such a remarkable occurrence that it was well worth putting on record.'

The first to respond was a bushman who had been in the Gulf of Carpentaria, which clearly had experienced earlier plagues, since Aborigines there were familiar with the rats, which they called *mungi*, and hollow trees occupied by owls and hawks were filled with rat bones, especially skulls. The bushman declared the irruption he had witnessed even more extraordinary than the stories heard by Price Fletcher. He knew his was the kind of account that 'many people would laugh at' as 'a thing belonging to legendary lore'. He vouched it was true by writing as 'One Who Has Seen And Felt Them'.

The bushman identified the rats as 'strictly nocturnal'. He recalled that they first appeared in 1869, when there was a big wet season with heavy floods in March causing great loss of sheep in the Gulf Country. While there were small numbers of rats around Christmas 1869, they increased rapidly in 1870 after another flood, remembered by some settlers as the biggest experienced by Europeans on the Flinders River. Rather than arriving in the Gulf Country from another part of the continent, the rats 'appeared to grow out of the ground beneath your feet, for their nests were often seen in the grass with young ones in them'. Before long, they 'covered the plains from the head of the Flinders River to its mouth'—the first account of the rats reaching the continent's north coast. They also extended 'south to Cooper Creek; and westward all over the plains and downs'.

Competition among the rats around the settlers' camps and stations was intense. They fought 'desperate battles over everything…standing on their hind legs, biting and squealing'. Flour particularly attracted them but the rats pursued any scrap of food or leather, occasionally eating the hobbles off horses. They were easily killed by bushmen who lured them with food, then gave them 'a tap with an iron ramrod or something weighty, until in an hour or two they would have a pile of perhaps a hundred or more'. But this killing had no more impact than predation by dingoes, owls and hawks, which were the rats' 'constant enemies'. If the men did not immediately bury the carcasses, other rats 'tried to drag away their dead for the purpose of eating them'.

This bushman remembered the plague continuing well into 1870. As the grass dried up that autumn, the rats 'took to eating whitewood saplings, cutting through with their sharp teeth sticks of 5 inches and 6 inches diameter'. They also found new homes in the cracks that opened in the clay soils. The rats inhabited these fissures, though they also accommodated vast numbers of snakes, which preyed on the rats, pursuing them from one crack to another. The rats began to decline only that winter when 'they seemed to sicken, and could scarcely crawl out of the way'. By the end of the year, they had disappeared, most likely due to 'their favourite food being supplied in too small quantities', and had not been seen again, because the rains since had not produced so much grass.

Another contributor, again a bushman, agreed that the 'great flood' in 1870 had resulted in an exceptional irruption of 'vermin of all sorts', with the rats a particular 'scourge' as they spread in 'hordes'. But he was eager to correct 'one particular—namely, that the rats did not completely disappear' that year. He reported that 'small families could always be found in the interior of the Flinders district', suggesting he was familiar with what scientists have come to call the rats' refuges, places where some are able to survive during dry periods, while those elsewhere die out. The locations of some of these refuges may vary from one dry period to another, while others appear to be enduring, with rats staying within them even during irruptions. The key to them is that they offer food and water and are relatively free of predation allowing the rat, which appears to live at most for about eighteen months in the wild, to breed in low numbers until good conditions return.

This bushman suggested that many colonists were puzzled by the appearance of the rats before the floods triggered the growth of native vegetation which he implicitly recognised as vital to their irruptions. He wondered how the rats 'lived, as at the time of their advent the rains had not been long enough to cause a good crop of grass or herbage, and this applied to the whole of the northern plains'. Some colonists believed that the rats were drawn 'instinctively…to country about to be visited by heavy rains; probably the atmosphere gave them the information long before the rains arrived from the regions of the north-west monsoon'. But the bushman dismissed this hypothesis because of the failure of the rats to return in plague proportions when there had been several more 'very wet and good seasons' later in the 1870s. Whereas the first contributor thought the rats had originated in the Gulf Country, the bushman endorsed the 'general belief' that the rats had travelled north from Cooper Creek where there also seems to have been big rains in 1869.

This plague also stuck in the memory of Edward Palmer, one of the first European settlers near the Gulf in 1864, who initially ran sheep, then cattle, on a vast station which a visiting journalist declared 'a magnificent run, almost as big one of the largest counties in England'. Much as Carl Feilberg discussed in his articles in the *Queenslander* that became *The Way We Civilise*, Palmer displayed no compunction about dispossessing Aboriginal people and killing Aboriginal men but, he emphasised, not women or children. In a letter to the Brisbane *Courier* in 1874, he declared that since 'no two races can inhabit the same country...the weaker must go to the wall'. Far from trying to hide the invaders' violence, he declared: 'Of course blacks are shot' to keep others in 'submission'. He acknowledged his own role, writing that he 'of course had collisions with blacks at times', though without any sense of it as murder.

Palmer softened his stance after achieving the dominion he considered his due. By the early 1880s, he was welcoming members of the local Mitakoodi and Mayi people on his station and had 'undertaken to protect them and give them a beast once a month or so—and let them have one side of the river to hunt on', with no concept of it being their land. He began expressing his admiration for their understanding of the country when he turned to writing about natural history and Aboriginal culture and, assisted by his wealth and standing, contributed to the Royal Society of New South Wales and the Anthropological Institute of Great Britain and Ireland. In 1885, after a Royal Society of Queensland was established in Brisbane, he addressed it on 'A Great Visitation of Rats in the North and North-Western Plain Country of Queensland'—the first paper about the long-haired rat to appear in an Australian scientific journal.

Palmer's account is largely consistent with those of Price Fletcher's contributors, but Palmer thought the plague had started already in mid-1869, when the rats moved north into the Gulf Country from the headwaters of the Flinders and Cloncurry rivers. To begin with, when the rats were few, most settlers did not know they were there. The colonists only realised when 'black-boys'—another of the settlers' demeaning terms for Aboriginal men, used by Palmer despite his new sympathy for them—went out on the plains and brought 'in scores of them, in an hour or two, for roasting as food'.

The catalyst for the rats reaching immense proportions was 'months of continuous rain' at the start of 1870, 'ending in the largest floods ever known', which in turn gave rise to an 'exuberance of vegetation', providing an abundance of food for the rat. As peabush sprang up all over the country, including many areas 'where it had not been known before', it grew eight and even ten feet high, and 'so close that it was almost impossible to make one's way through it'. The impact of the rat was manifest along the Flinders and its tributaries for hundreds of miles as 'the grass looked as if it had been cut down, or flocks of sheep had been over it'. The rat's favourite resort was large, open plains. 'Fifty thousand square miles occupied by these animals, and one rat to every ten square yards in each mile would not represent anything like their numbers.' In other words, Palmer figured there must have been at least 150 million rats.

Palmer recalled the plague declining towards the end of 1870 and ending in 1871. One cause was the dryness of the grass. Another was the increase of the rat's natural enemies—hawks, owls, dingoes and brown snakes 'very numerous…to be met with any day on the plains in scores'. A third was the rat's voracity, 'leading them to devour each other at all times and more particularly when pressed by hunger'. Had the rat 'continued to multiply or even maintain their number, the country would have been uninhabitable'.

Unlike the *Queenslander*'s contributors, Palmer attempted a detailed description of the rat—likening it to the brown rat, now known as *Rattus norvegicus*, in both its habits and appearance. It was 'given to burrowing' where it could not 'appropriate any burrows or cracks already made', was 'greyish brown', with its fur 'close and short', body 'thick and strong', ears 'short and stiff'. Although Palmer remembered its body as 'not much more than six or seven inches long', he thought its tail was just three inches when it would have been almost as long as its body. He believed the rat was not 'a permanent resident', perhaps because his station did not include any of its refuges, perhaps because he remained oblivious to them.

Missing from these accounts was an irruption much further south. The rats were recorded in January 1871 by Sub-inspector James Gilmour of Queensland's Native Police based on the Bulloo. While travelling in the catchment of Cooper Creek searching for traces of Ludwig Leichhardt who had disappeared on his second major expedition more than twenty years earlier, Gilmour encountered 'large quantities' of rats. From April, rats excited extensive attention in South Australia, where a correspondent at Mount Margaret had experienced an 'excessively annoying' plague in 1868, but the 'great northern army' of rats was new to most settlers.

A pastoralist on South Australia's Western Plains reported that the land was 'overrun', with the rats 'not in hundreds or thousands but millions'. To evoke their multitude, he described how they 'obliterate the tracks of sheep and cattle. The tracks or sheep pads that my flock make in coming to the yards are not to be seen in the morning from the number of rats that have been running over them in the night.' In doing so, he implicitly corroborated William Brahe's evidence to the Burke and Wills inquiry, confirming how easily the rats could have removed all traces of the explorers at their depot on Cooper Creek in 1861.

Another report came from Lake Koppermana, the site of South

Australia's most northerly police station staffed by Samuel Gason of the South Australian Mounted Police who claimed to have been accepted by the Diyari as one of them and to know more about their culture than any other colonist due to his command of their language. When Gason published a book about the Diyari, he included the *miaroo* among seven species of native rat and mouse, identifying it as one of the Diyari's totems. The report from Lake Koppermana in June 1871 stated that the rat was 'increasing'. The police had killed six hundred in six weeks, which became six hundred in a night as the story circulated.

Several more reports came from one of the *South Australian Register*'s contributors following the Overland Telegraph, which was under construction. As this contributor headed north, he encountered the rat 'more or less all the way' from Boorloo Springs to Finniss Springs, a distance of about sixty miles, then encountered more in other places. He reported that colonists were acquiring kittens with a view to raising sufficient cats to eat all the rats in the north. Unlike his predecessors, he assumed public knowledge of these 'vermin', identifying them as 'heard so often about'.

The Flinders Ranges were another domain of the rat. *Maiurru Mitha Vambata*, meaning 'dirt dug out of a hole by rat', was the name of the local Adnyamathanha people for the low hill that Edward John Eyre miscast as Mount Hopeless. The rat featured too in a Dreaming story of the Adnyamathanha, suggesting they hunted the rat by plugging its burrows' exits before digging it out.

When the rats appeared in 1871, Adnyamathanha people told George Debney of Mundowadana Station that *my-ar-roo* seldom visited, 'the latest instance having been many years ago, and before the whites settled in the Far North'. The Adnyamathanha identified the rat as 'coming down in myriads' from the east and north-east where floods were 'driving them down'. Edward John Eyre had already heard of this phenomenon in the 1840s when he wrote of Aborigines procuring rats 'in the greatest numbers and with the utmost facility when the approach of the floods in the river flats compels them to evacuate their domiciles'. Debney recorded that the Diyari considered the rat's 'visitation as a windfall' and regarded them as 'a great dainty'.

Another account, from Hookina in the Flinders Ranges, was of 'swarms of rats' eating 'paper, books—in fact, almost anything they can get at', and 'falling down the wells, making the water undrinkable until the wells be cleared out'. The most significant reports, in terms of the rat's range, were that it was to be found around Port Augusta on Spencer Gulf, to the south-east at Melrose, and was 'slowly but surely marching' even further south. These accounts are consistent with bones of the rat being found in the pellets of owls not all that far away at Burra Creek and Mount Bryan.

For the next two years, the rats abounded in at least a few parts of South Australia. In 1872, a colonist at Mount Margaret thought they were 'rightly termed a plague, for nothing is sacred to their teeth and claws, and they have a particular weakness for wearing apparel'. He hoped they would 'seek new pastures before long' and 'not take possession of the place like they tried to do on a former occasion'. When the rats again came down 'in countless hosts' to the Western Plains in 1873, 'the settlers at first tried to hold their own against the invaders, but the combat was useless, for the immense numbers killed seemed not in the least to have lessened the mighty crowds that still pressed onward'. Aboriginal people, meanwhile, welcomed the rats. Like the Diyari, the Arabana people west of Strangways Springs considered them a 'blessing'.

CHAPTER 6

The Night Kite

Birds enlivened Kenric Harold Bennett's dull days as a pastoral worker. He would be on the plains, far from trees, with a flock of sheep, profoundly bored, hoping vainly for something of interest. Then his sheep would disturb a red quail in the grass, there would be a rushing overhead and, to his awe and delight, he would see a black falcon descending with 'ferocious rapidity'. Thirty or forty feet from the ground, it would arrest its descent, chase and kill the quail, then ascend 'by a series of graceful curves', devouring its prey, while 'an occasional tuft of blood-stained feathers slowly wafted earthwards, evidencing its success'.

Bennett's close observation of such scenes, and precise memories of them, extended to the birds that preyed on the rats back of the Lachlan in 1864. Fifteen years later, having seen these birds again only in 1870, when a few appeared as introduced mice plagued the back country, Bennett described them exactly in the *Queenslander*: 'The plumage…on the breast and under parts was pure white, the back—with the exception of the shoulder coverts—light grey, almost white, the shoulder coverts black'. But Bennett did not know the identity of this bird, which he thought of as a 'hawk'. Nor did he realise that, though Europeans had, episodically, been learning about it for forty years, he knew most about it and its hunting of the long-haired rat.

It entered western science in 1842 when John Gould identified one from South Australia as an *Elanus* or kite. He called it *Elanus scriptus* because the undersides of its white wings appeared to have black script on them. Two years later, as Charles Sturt headed inland towards the hills he named the Barrier Ranges, which would become the site of Broken Hill, he sighted one of these 'letter-winged kites'. Soon he could see flocks of thirty to forty—evidence these kites were exceptionally gregarious for birds of prey. Over the following year, Sturt's party continued to see these birds, sometimes congregated in trees, sometimes soaring through the air or hovering, but Sturt never saw them take prey on the ground. He and his men could not determine what they ate though they were 'led to believe'—presumably when they asked Aboriginal people—that the kites fed on mice.

Samuel White, a South Australian pastoralist whose passion for ornithology eclipsed his interest in managing his sheep station in the colony's south-east, discovered more in 1863. That July, at his own expense and with just observing and collecting in mind, the twenty-eight-year-old White headed north with a dray, three horses, ten months' supplies and a servant called Cottrell. Four months later, White and Cottrell reached Lake Hope, which was full after heavy rain further inland. There, to his immense excitement, White became the first settler to observe letter-winged kites closely. Most likely, he continued to encounter the kites as he proceeded along Cooper Creek, where the last of his horses died, forcing him to abandon his dray and his specimens, except those most precious to him. They included one kite skin and two kite eggs that Cottrell and he carried back when they returned south on foot.

A museum had just opened in Adelaide. Its curator, Frederick Waterhouse, whose brother George first applied the name *Mus longipilis*, was eager to build a collection. But White sent his observations and specimens to London where John Gould, in a marked departure from his lavishly-illustrated folios, was looking to reach a general audience through a two-volume, unillustrated *Handbook to the Birds of Australia*, something he never did with his work on mammals. Gould included White's new material in this *Handbook* in 1865, purporting to quote him at length but slightly rewriting his account. White reported that the kites did not just congregate together but bred in colonies comprising as many as thirty nesting pairs. 'It flies when near the ground with a heavy flapping motion, but occasionally will soar very very high when its movements are very graceful'. Its nests were 'composed of sticks, lined with the pellets ejected from their stomachs', as the kite regurgitated material it found indigestible. White identified it for the first time as primarily subsisting on rats, not mice. It must have abounded at Lake Hope and on the Cooper when he was there in 1863, as it did on the Flinders River. The kites' pellets were 'principally composed of the fur of the rats'.

Sturt and White implied the kites were diurnal, active only by day. Sylvester Diggles, an artist, musician and naturalist, who produced a twenty-one-part illustrated *Ornithology of Australia*, concurred in 1875 in an article in the *Queenslander*. The bushman, who wrote to the *Queenslander* in 1879 as 'One Who Has Seen And Felt Them' after Price Fletcher asked for accounts of rat plagues, suggested otherwise. He identified these birds as 'night hawks', indicating they were nocturnal.

Kenric Harold Bennett provided the most detailed account in the *Queenslander* a few months later. Bennett recalled how these 'hawks' spent each day sitting on the branches of dead callitris pine trees. They sat so closely packed that, 'at a distance, and with their breasts turned towards the observer, the branches looked as if covered with snow'. They typically were so sluggish that Bennett had seen Aboriginal boys 'knocking them off on one side of a tree with their boomerangs, while those on the other side remained perfectly still...apparently quite unconscious of the danger'. But in the late afternoon they all took to the wing, rose by a series of graceful circles 'until their white breasts became mere specks in the blue sky', then descended in the same manner and resumed their perches. At dusk, they left again to hunt the rats. They would 'fly out over the plain, over which they might be observed hovering like sea-gulls, and every now and then pouncing down upon a rodent, which they would devour whilst on the wing'. By sunrise next morning, the birds would all have returned to their roosts and the ground beneath would be thick with balls of rat fur.

These observations were momentous—and not just because all that Bennett reported was new. The letter-winged kite is an ornithological oddity. Every other eagle, hawk, falcon and kite, now known to number almost three hundred species, is diurnal. Only the letter-winged kite almost always hunts at night. But Bennett, typically, lacked the broader ornithological understanding to draw out the significance of his observations when he wrote to the *Queenslander*.

Contributing to it not only taught Bennett much but emboldened him to explore other forums for revealing what he had observed and collected. He started with the Sydney International Exhibition of 1879—Australia's first world's fair—assisted by his uncle William Brodribb, who was one of the exhibition's commissioners. He continued by offering specimens to the Australian Museum, which put him in touch with its curator, Edward Pierson Ramsay, who provided him with scientific literature enabling him to learn that he had seen the letter-winged kite and to understand the significance of his observations of its nocturnal hunting of the rats.

Bennett's contributions to the Sydney International Exhibition conventionally included an array of animals he had killed: stuffed birds; kangaroo, wallaby, possum, dingo and bullock skins; a nine-foot-long snake; and a rug with 'a great variety of parrot skins' in it. He attracted much more attention by contributing live animals, something usually done only for introduced species. One was a mallee hen, which Bennet may have raised. The others were bilbies which, because of their small bodies and very long ears, some colonists dubbed 'rock rabbits' or 'rabbit-eared bandicoots'. Bennett followed the Yuwaalaraay people of north-western New South Wales and south-western Queensland in calling them 'bilbahs'. Those collected by Bennett would become known as the greater bilby, *Macrotis lagotis*, while a smaller species found in the deserts of central Australia became *Macrotis leucura*.

When Gerard Krefft had been at the junction of the Murray and Darling rivers in 1856–57, he found that, like the greater stick-nest rat, the greater bilby was in retreat, no longer to be found in northern Victoria. But it remained in western New South Wales where the journalist-drover who visited Yandembah and Moolah in 1877 reported that the 'wonderful little bilby' was to be found 'all over' the back country in 'almost incredible numbers'. Because its burrow has only one opening, Bennett would plug it, then try to dig out the bilby, only to find, as he described with exasperation, that often the bilby rapidly created a new exit and ran off, several feet from where he was 'toiling in a state of profuse perspiration'. Because the two specimens he sent to the 1879 exhibition were the first to go on show in Sydney still living, they were identified as 'special', even 'unique', and featured in advertisements promoting the exhibition, despite responding to their display in daylight by 'coiling up against each other in such a manner that very little of their form could be seen'. Since they were alive, they were placed in the stock shed of the exhibition's agricultural hall—included bizarrely in the 'Poultry, Pigeons' category, and winning Bennett its first prize.

Bennett was also lauded for a book containing ninety specimens of grasses which, most likely with Brodribb as intermediary, were identified by another of the great German figures of Australian science, the Victorian Government Botanist, Baron Ferdinand von Mueller. The *Australian Town and Country Journal* published a letter from New South Wales Government Botanist Charles Moore, revealing that von Mueller and Moore were 'greatly pleased' with this 'fine collection…so carefully and beautifully preserved' by Bennett, 'a man of scientific tastes, whose acquaintance is well worth cultivating'. The *Journal* itself considered Bennett's book a model to 'be looked at, and carefully studied by all our young people in the colony'. The *Goulburn Herald* agreed. 'Our bush is wearisome and monotonous only to those who pass through it without notice and without interest', it declared. 'To the observing and inquiring, new wonders will appear at every step.'

Like many contributors to the exhibition, Bennett was also

corrected, admonished and denigrated by critics eager to display their superior knowledge. One of his other exhibits was a case which he identified as containing eighty sorts of wood from the saltbush country between the Lachlan and the Darling rivers—implicitly suggesting the great diversity of this country so often dismissed as barren. Bennett was instructed that this collection contained many duplicates and the samples were 'useless for educational purposes' because they were not named. He was also advised that his book of grasses would have been more useful had he recorded where and when he collected his specimens. A card table inlaid, most likely by Bennett himself, with 500 pieces of his eighty sorts of wood, attracted no notice because it was competing with other inlaid work containing up to 8000 pieces.

Bennett learned from these criticisms, but he learned much more by making contact with the Australian Museum where the trustees had fallen out with Gerard Krefft, partly because he embraced Darwin's theory of evolution. After employing two prize-fighters to carry Krefft onto the street in his curatorial chair, the trustees replaced him with Edward Pierson Ramsay, the museum's first Australian-born curator, who enjoyed exceptional scientific opportunities in Sydney due to a substantial inheritance and significant patronage. Bennett was delighted when he first offered birds to the museum as a gift in 1879 and Ramsay responded with a 'highly complimentary letter', accepting most. 'I feel quite proud that my humble endeavours have proved of such interest and that many of the objects I have collected are deemed worthy of a place in such an institution as your museum', Bennett replied in turn.

Ramsay responded so enthusiastically partly because of the prospect of securing even more material. High on his wish list were human remains that had featured prominently in the museum's display since the mid-1860s when the first showcase in its main hall included several Aboriginal skulls—considered to be 'of very low formation' or sub-human. Like other curators in Australia, Ramsay wanted more of these 'relics', ideally 'perfect skulls or skeletons', which he declared 'very valuable'.

Bennett, who had never opened graves, knew that an Aboriginal burial ground was immediately accessible a hundred yards from Moolah's homestead where he lived. But, as he explained to Ramsay, he did not want to handle 'these evidences of mortality'. Nor, with many Aboriginal people living on Moolah, did he want to 'displease' the descendants of those whom colonists considered 'nameless dead'. Bennett identified the opening of Aboriginal graves as desecration. Still, he complied immediately to curry favour with Ramsay. 'Anything I can do to assist you I shall be happy to do', he declared. Before long, he had opened about a dozen graves which he estimated were up to forty years old. While most of the remains 'crumbled at the touch', the bones in one grave were in 'wonderful preservation', leading Bennett to box them up and send them to Sydney 'for the cause of science'.

His need for basic ornithological instruction from Ramsay was patent when the museum formally acknowledged his first donation of bird skins, identifying those it had accepted with the birds' Latin names, which were all meaningless to Bennett. What he needed first, Bennett admitted to Ramsay, was the birds' common names. If Ramsay could tell him which was a pigeon, parrot, duck, quail and so on, he would find it of 'great use' and so it was, though these names were only a start. As Bennett revealed in the *Queenslander* in 1880, he had skins of more than seventy birds from the back country, and still did not know the English name of most of them.

One of the scientific publications Ramsay sent Bennett was Ramsay's first book—the opening volume of his *Catalogue of the Australian Birds in the Australian Museum*, which focused on the *Accipitres* or diurnal birds of prey, a favourite of ornithologists. When Bennett found *Elanus scriptus* in this catalogue, he immediately recognised it as the bird he had seen preying on the rats back of the Lachlan in 1864. Ramsay's description corresponded with his memory of the bird, as did Ramsay's characterisation of it as the 'letter-winged kite' and 'letter-winged mouse-hawk'. At last, Bennett knew this bird's popular names as well as its Latinate designation. But Ramsay's identification of the kite as diurnal was contrary to Bennett's experience. 'I think you are in error,' Bennett advised Ramsay. 'Although it is to be seen at all hours during the day perched upon some branch still so far as my experience goes (and I am well acquainted with its habits) I never knew it to seek its prey (rats and mice) until night and I have often seen them on moonlit nights catching the above-named rodents.'

Ramsay did not expect his bush informants to address him in this fashion: they were to provide specimens, not challenge his published work. Although he had never seen a letter-winged kite in the field, Ramsay remained confident it was diurnal because that was the case with every other kite, hawk, falcon and eagle. If Ramsay checked the specimens in the museum's collection for any signs that Bennett was correct, the only one he may have noticed was their soft plumage, characteristic of nocturnal hunters, enabling them to come on their prey silently. While the eyes of the letter-winged kites are forward-facing, black-ringed and very large like those of owls, they are only slightly bigger than those of the closely related black-shouldered kite, which is diurnal.

Ramsay responded by interrogating Bennett about his observations, manifestly sceptical, only for Bennett to invoke his field experience, which Ramsay lacked. 'Since I have resided in this portion of New South Wales (18 years)', Bennett reiterated, 'I am *perfectly certain* as to the nocturnal habits of *E. scriptus*', now happily using its Latin name. 'I have seen scores of them at all hours during moonlit nights catching rats and I am equally *certain* that when the birds left this camp at dusk they *did not* return until just as the sun was rising the next morning and of this I had ample proof for many months.' He enclosed what he had written in the *Queenslander* in 1879, perhaps thinking his published account would be more persuasive. 'It may be of interest to you,' Bennett wrote. 'Please return it when you have done with it,' he requested, suggesting it was his only copy.

Scientists in London often mistreated their colonial associates—appropriating their discoveries, stopping them publishing or ignoring their observations because they did not fit their preconceptions. At most, they gave the colonists some acknowledgment. Government scientists in the colonial capitals treated bush naturalists similarly. Bennett was insulted when Ramsay, wary of frauds, questioned the authenticity of some of his specimens. 'What a sceptic you are', exclaimed Bennett, explaining that he only sent eggs that he himself had collected. He complained at Ramsay's 'caesar-like' tone. He was vexed that, while he answered Ramsay's letters immediately, Ramsay, who had many more correspondents, was much slower and occasionally did not respond. 'Not a scratch of a pen from you! You are a pretty sort of a correspondent are you not? Does not your conscience prick you?' Bennett once wrote. Apparently it did not, which caused Bennett again to write: 'Not a scratch of a pen from you.

Aren't you ashamed of yourself?' Yet this disappointment was also an expression of how much Bennett valued their relationship. 'I should so much like to see you and have a long yarn with you', he declared, identifying Ramsay as one of his 'best friends'.

Bennett's identification of the letter-winged kite as nocturnal was an obvious subject when he graduated from the *Queenslander* to more prestigious scientific writing. His prime forum from 1881 was the Linnean Society of New South Wales, with Ramsay his conduit as the society's Honorary Secretary. Bennett relished this opportunity, though his distance from Sydney meant he was generally unable to present his papers in person, depriving him of the opportunity to take part in the discussion that followed, and in at least one case leaving him fuming at what he regarded as unsubstantiated criticism of his findings. He had six papers published in the society's journal and became one of its fellows, published four more in the journal of the Royal Society of New South Wales and another in the Australian Museum's *Records*.

The first was typically about birds and exemplified Bennett's capacity to inform and elucidate. John Gould had recorded that the black-breasted buzzard, one of Australia's largest raptors, would drive an emu from a nest, pick up a stone with its feet and, while hovering over the nest, drop the stone to break the eggs, then devour them. But Gould had published this account 'without vouching for its truth', both because it seemed so extraordinary and because its source was 'the testimony of the natives'. When Bennett heard this story from Aboriginal people from across the plains bordering the Murrumbidgee and Lachlan rivers, he too was 'inclined to disbelieve'. Then one of Bennett's friends partially corroborated it as did Bennett himself without realising he had explored what Gould wanted clarified, since he had not read Gould's work.

His one paper about Aboriginal occupation of the land explained how 'back country natives' in the arid terrain between the Lachlan and Darling survived by obtaining water from the roots of eucalypt, hakea and kurrajong trees. Using wooden shovels, they removed the soil above these roots for a distance of twenty or thirty feet, cut the roots into small sections, placed them on end, and drank a water which was 'beautifully clear, cool, and free from any unpleasant taste or smell'. As proof, he sent a bottle to another of his new scientific contacts in Sydney, the curator of its Technological Museum, J. H. Maiden, who testified to the water's 'good quality'.

Bennett did not devote a paper to the letter-winged kite and its hunting of the rat because, even if Ramsay did not censor his contributions to the Linnean Society, their association was too important to Bennett for him to embarrass Ramsay by contradicting him. A visit to Sydney by Bennett in 1881 was transformative of their relationship as well as a much anticipated, treasured opportunity for Bennett to examine specimens and books unavailable to him. When they resumed their correspondence, Bennett continued opening his letters, 'My dear Ramsay', but within them sometimes addressed him with new familiarity as 'Old Man'.

One of Bennett's great desires was Gould's two-volume, unillustrated popularisation of his work on birds. In 1881, Bennett informed Ramsay that he was eager to get this handbook and asked its cost. In 1885, having received nothing from Ramsay for more than five years but having given him many birds and eggs for his private collection, Bennett reminded Ramsay that he still sought the handbook. When Ramsay gave him it a few months later, Bennett responded: 'You could not have thought of a more acceptable present.' Its entry on the letter-winged kite was one of the first Bennett read. In his letter of thanks, Bennett wrote that he was 'under the impression' that it was the bird he had identified as nocturnal. He advised Ramsay that the question could easily be decided by examining specimens he had sent to Sydney at different times.

Bennett wrote about the letter-winged kite for the Linnean Society a few months later, but buried his account within a section on the black falcon in a paper titled 'Birds breeding in the Interior of New South Wales'. He did so by including an extract from his letter to the *Queenslander* in 1879. His one advance on what he wrote there was to identify its 'hawks' as *Elanus scriptus*. But while he republished the passage describing how the kites sat motionless most of the day, then hunted the rats at night, he did not draw out that the kites were nocturnal, let alone identify *Elanus scriptus* as the one nocturnal raptor. He did not make it explicit that Ramsay had classified the kite incorrectly. When Ramsay published a supplement to his *Catalogue* in 1885, he ignored Bennett's observations—he still characterised the kite as diurnal and failed to mention that it preyed on rats.

CHAPTER 7

City of Iron

Winton in western Queensland was 'a pretty little city in embryo' in 1880, its buildings easily itemised by those writing about it. Two years after being established on the fringe of the terrain colonists called the Never-Never, this 'furthest town out west' comprised three hotels, three stores, two boarding houses, two butchers, two billiard rooms, a police station, a post office, a blacksmith, a bootmaker, a saddler and ten houses. It also was home to the Never-Never Amateur Jockey Club, whose twice-yearly meetings featured steeplechase, hurdles and flat racing.

No one counted Winton's settlers, who probably numbered fewer than a hundred, though many more visited for the biennial races. Nor did anyone count Winton's Aboriginal population, whose camp was within sight of the colonists' settlement but separate from it. The long-haired rats were different in the wake of vast rains late in 1879, followed by more in the spring of 1880, that saw waterholes overflow, rivers flood and grass grow everywhere. When the rats appeared that May in Winton and stayed three months, colonists recorded the exact tally killed one night, as part of their obsession with the rat's numbers.

The rat was common in western Queensland in 1879, several colonists advised Price Fletcher of the *Queenslander*, and such reports intensified in other newspapers in early 1880. A prospector, bound for the Mackinlay Ranges, crossed the swollen Diamantina River in the hope of escaping rats 'playing the deuce' with his party's rations and saddles. A surveyor on the Herbert River reported that rats were 'in thousands' and 'swarmed about his stores'. The plains around Cloncurry were 'perfectly inundated', so it was again 'quite impossible to leave on the ground anything in the shape of rations or saddlery for a single night without finding them considerably damaged in the morning'.

Much the same was true of South Australia's far north. In January 1879, the surveyor W.H. Cornish recorded meeting a party of about 100 Queensland Aborigines who 'had been driven down by the black police' after killing 'a great many sheep and a shepherd'. Having visited one of Cornish's camps, they went off to get *miaroo*, which Cornish identified as 'their chief animal food'. A year later, Cornish observed that, 'if it were not for the rats, which are in thousands and persistently endeavour to destroy everything which comes up', there would be 'a good show of vegetables' in the settlers' gardens. Before long, Cornish was killing more than thirty rats a night.

Another account came in July 1880 from the Warburton River, to the north-east of Lake Eyre. According to a local settler, the country was 'quite ploughed up with rat-holes'. His focus was the different responses of Aborigines and colonists. 'The blacks rejoice and eat, and eat until they are sick and then commence again', he wrote. 'This kind of living seems to suit the natives for they look fat and shiny just now.' Settlers felt besieged. One 'poor fellow' was riding about with a hat that had little more than the brim left, 'the rats having made short work of the crown'.

Another report came from Edward B. Sanger, a geologist whose broad-ranging scientific and literary interests led him to join the Adelaide Philosophical Society, become a corresponding member of the Linnean Society of New South Wales and contribute to Melbourne's *Victorian Review*. After fleeing Adelaide to escape bankruptcy, Sanger published one of the best-informed accounts of the rats in the *American Naturalist*. He described them as infesting 'large tracts of country in droves during flood time. They migrate from place to place. Their well-beaten paths may often be seen winding through the sandhills, and sometimes the droves themselves'. He too identified them as *miaroo*.

Kenric Harold Bennett was among a small group of colonists with a sharp eye for the settlers' impacts on the land. He observed that several species of bird, plentiful when he arrived back of the Lachlan, had disappeared and were 'only found in thick timbered or scrubby country some 60 miles to the north'. Some land, treeless in 1862, had become thick with the native callitris pine, which he attributed to exceptionally wet years, such as 1870, rather than the end of regular Aboriginal burning of the country. Elsewhere, vast tracts of eucalypts had died. Bennett blamed their deaths on a proliferation of possums, which had stripped the trees of their foliage. In turn, he attributed the possums' numbers to Aborigines no longer hunting them.

Price Fletcher was particularly concerned for the emu. He

thought it in such decline on Queensland's western plains that it would soon be extinct, prompting him to pen 'a plea for protection of our feathered giants'. But he reconsidered after some of his readers advised that it was on the increase elsewhere. Bennett reported that the emu was 'extremely numerous' back of the Lachlan. Because it damaged wire fences and frightened ewes with lambs, Bennett's father had some of his station hands destroy emu eggs, smashing 132 in one paddock in a day. Within a few years, the emu was in decline there too as it was 'hunted down with kangaroo dogs, shot down with a rifle or fowling piece', and its eggs continued to be 'ruthlessly destroyed'.

The dingo—the largest terrestrial predator on the continent, following the extinction of the mainland thylacine—killed vast numbers of sheep. Colonists responded by trapping, hunting and poisoning it. The New South Wales parliament legislated specifically against it through 'An Act to facilitate and encourage the destruction of native dogs'. As strychnine proved effective, but killed many other creatures, 'dingo poisoner' became a new colonial occupation, with the 'poison cart' its conveyance. Settlers succeeded in eradicating dingoes from many areas, but then their traditional prey began to multiply, creating new problems for pastoralists. 'Now we have kangaroo revelling in thousands where only dozens of dingoes lived', ran a typical account.

Bennett was a keen observer, participant and commentator. He recorded how, in less than a decade, his father transformed the dingo from 'swarming' on Yandembah to 'a thing of the past'. When William Brodribb first stocked Moolah with 5000 ewes, dingoes caused the deaths of 1000, partly by rushing them through fences into waterless paddocks where they died of thirst, fueling Brodribb's decision to stock Moolah with cattle since they were not as vulnerable. The journalist-drover, who visited in 1877, found the tails of sixteen dingoes, all killed within a week, on display. While some colonists came to lament that they had upset 'the balance of nature' by killing so many of these dogs, Bennett still thought the dingo such a 'curse' and 'brute' that he hoped for its extermination.

Colonisation was also transforming the irruptions of the long-haired rat not only because of the settlers' destruction of the dingo—one of the rat's main predators—but also because of the attractions of the settlers' abundant food and equipment. Winton, the first town reached by the rat, proved to be a huge lure, like a giant explorers' camp. The local contributor to Rockhampton's *Northern Argus* reported in May 1880 that a 'perfect deluge of rats' had 'swept down upon the town from all directions across the plains'. 'How are you off for rats?' the local correspondent of the *Brisbane Courier* asked. 'We can spare you 1,498,765, and then have plenty to spare.'

One of the most famous rat tales—the story of the town and country rats in La Fontaine's adaptation of *Aesop's Fables*—provided a comparison. 'They do not hide in holes and galleries like La Fontaine's rat de ville but rush about in open daylight in scorn of cats, dogs, traps and sticks,' the correspondent of the *Northern Argus* wrote improbably, given the rats were generally nocturnal. 'There are rats of every size, rank, and denomination, from the grey-headed old general, the leader of a battalion, to the young recruit...They invade every place, respecting no thing and no person. Conscious of their numerical strength they seize upon all before them that has the least appearance of being edible, and their tracks are marked by ruin and destruction to the goods and chattels of everyone in the place.' In sum, there was 'no withstanding them'.

Winton's main buildings were corrugated iron—a prime outback material because it was light, strong and easily transported. These buildings included the new police station—erected after local residents initially depended on a 'vigilance committee' to safeguard themselves and their properties when 'at frequent intervals, gangs of bush larrikins, engaged on the neighbouring stations, would come to the town'. Just fifteen by eight feet, this all-metal station was like an 'oven' on summer days, as frightful for the two officers who lived there as for those held in its lock-up. With the rats devouring everything from 'socks to doors', the *Northern Argus* declared it fortunate that Winton was 'a city of iron', so there was 'a chance at least of the buildings being left standing'.

Local residents tried many means of control. 'We knock them over with sticks, staves and stones: we secure them in pitfalls, kick them out of our road and trample on them; but without making any apparent diminution on their numbers, and certainly without intimidating the survivors'. Mr Fox, one of the partners in a local carrying company, was seen 'literally staggering of a morning under the weight of a large wheel-barrow-full of dead rats destroyed during the night in his store'. The Wintonians, or Wintonites, as they were also known, could 'keep nothing from their ravages', and there was 'scarcely a man in town' whose boots had 'not fallen a prey to the invaders'.

The incident that attracted most attention occurred at one of the town's 'grand establishments'—the iron-roofed, iron-walled Corfield and Fitzmaurice Hotel, boasting a bar, dining room, billiard room, coffee room and eleven bedrooms. A travelling salesman would recall that, when he stayed in one of them, he passed 'a night of horror' as 'rats in thousands' bit his toes, nibbled his hair and ran over his face. In response, as reported in May

1880, 'The cook at Fitzmaurice's hotel caught in one night 142; this was for a bet the cook won, and would again if anyone was game to wager, which no one is.' As Corfield recalled it years later, a wheelbarrow again provided a receptacle and a measure: 'I told the Chinaman cook of the hotel that I would give him a pound of tobacco if he caught a hundred rats. That night...I was several times awakened by what seemed to be stamping of feet. In the morning I found that the Chinaman had obtained an ironbark wooden shutter and rigged up a figure-four trap with bait underneath, and by this means obtained a wheelbarrow full of dead rats.'

The cook's name was Ching Foo. The '*chef de cuisine*', he was mockingly dubbed, as Anglo-Irish colonists derided the Chinese. 'If I could get a sober, reliable white man to cook in my hotel at the very good wages that I give the Chinaman, I would be only too glad', a Winton publican wrote a few years later. Only the Chinese would cook 'in quiet times, at race time, and Christmas time'. Ching Foo's rat-killing was considered just 'an ordinary night's work for this expert Celestial'. The implication was that Ching Foo was an old hand at rat-catching; colonists assumed that Chinese considered rats 'number one good eating'. But this assumption was another product of westerners' antipathy to Chinese and denigration of their culture. Apart from those from Canton, Chinese were generally as loath to eat rats as Europeans were. Most likely, Ching Foo did not eat them, either in China or in Winton. He certainly did not put them on the menu at the Corfield and Fitzmaurice Hotel. Otherwise, he—and they—would have become even more of a story.

That July came another report of the rats being carnivorous but finding new prey—'playing mischief with the green lambs at the lambing, fifteen and twenty rats attacking the lambs and eating them before they are strong enough to help themselves'. But by late August, the rats were almost gone. 'Just fancy', a local resident observed, 'one month you could count them by millions, and in another there are hardly any to be found.' Colonists identified one cause as 'water drying up'. Another was dingoes 'gathering where any water was left, and...devouring all the rats in the immediate neighbourhood'. Early the following year, a journalist on the Diamantina thought of the dingo as eking out 'a precarious living in the shape of pigeons, lizards and the larger insects', except 'during the rat seasons'.

Price Fletcher at the *Queenslander* was not interested. 'An invading army of rats has been devastating the Upper Diamantina', was all he wrote. His new ambition was that readers provide monthly notes, which would 'let us into all the secrets of our Australian natural history' because they would be 'valuable as facts', 'recorded at the time', rather than written up much later from memory. 'We want our bushmen to turn observers', he declared. 'They are at present, with very few exceptions, like a man in a garden of flowers who deliberately shuts his eyes to the beauties around him.'

That August, a reader on the Landsborough River in the Gulf Country reported: 'The white owls have followed the rats, which have nearly all left.' Kenric Harold Bennett, Fletcher's exemplary contributor of monthly notes, began by recording that a single 'letter-winged mouse-hawk' had reached Yandembah, the first of these kites he had seen since 1870. When more arrived, he shot four with a single shot as they typically roosted very close to each other. When he examined their stomachs, he 'could not detect the slightest trace of food of any kind'. Most likely, they were birds that had feasted on the rats during the irruption in Queensland, then dispersed at the end of the plague in search of food but, back of the Lachlan, were finding none.

Another 'army' soon appeared around Winton, this time venomous snakes—'a swarm of serpents' that feasted on the rats and attacked other animals. One station was said 'to have lost £200 in horseflesh'. That report was soon transformed through confusion or hyperbole into a story that the snakes would kill '200-guinea horses in twenty minutes'. That sparked stories that not only the snakes but also the rats were poisonous. Just one of their bites could be fatal, it was suggested, because of the rats' venom. 'A lamb bitten by a rat seldom, if ever, recovers', ran one account.

Looking back on this irruption, a correspondent of the *Aramac Mail* wrote: 'They came in droves from some unknown locality further out. Flour, sugar and other consumable goods had to be stacked on platforms proof against their climbing capabilities; the gardens were eaten completely out.' He found another literary analogy in Bishop Hatto who, legend had it, was devoured by a myriad of rats at the end of the tenth century, even though the bishop's castle was perched on a rock in the Rhine River. Colonial reality was far more terrifying than medieval folklore, this journalist suggested. The 'army of rats' invading Winton had been 'so numerous that the company which devoured Bishop Hatto were nothing in comparison'.

Carl Lumholtz, a Norwegian zoological collector, heard more stories in 1881 while visiting Vindex Station outside Winton. Lumholtz was told that, rather than coming from all directions, the rats moved en masse, approaching 'from the north-west and proceeded via Winton, on their wanderings towards the east'. A man on another station recounted that 'one night for amusement he laid a piece of meat on his threshold, and killed with a stick 400 of these animals which came up to eat the meat'. Lumholtz later learned that 'an army of rats…greatly reduced in number', had also passed the township of Westwood, about 500 miles from Winton, and thirty miles from Rockhampton, the first suggestion of the long-haired rat nearing Australia's east coast. Lumholtz thought these rats were 'doubtless the same clan' as those that had been around Winton. But given the distance and the dearth of other reports of the rats travelling so far east in late 1880 or early 1881, Lumholtz may have been mistaken.

More tales waited to be recorded. A reporter with the Brisbane *Telegraph* who visited Vindex in 1883 wrote of 'strange, marvellous legends repeated with gusto to the lonely wayfarer', as if what he was told about the rats was not believable. Yet most of what he reported was consistent with accounts from 1880 and probably occurred. He wrote: 'The rats appeared in thousands, destroyed all the vegetables in a garden on the back of the creek, hamstrung young lambs just after they were dropped, ate the hair off people's heads while they were asleep, consumed boots while the owners of them slumbered unconscious of the depredations being committed at their bedside, ate or destroyed saddler or harnesses, and then vanished.'

Confusion and hyperbole grew over time. While long-haired rats have lightly haired tails typical of rodents, one account claimed that they had 'a tuft of hair on the tip of the tail', others that they were 'bushy-tailed'. They were said to have come from the south and headed to the Gulf of Carpentaria; to have come from the north-west and headed south-east; to have 'travelled in mass formation, covering the country 200 to 300 miles wide…in a solid phalanx for a few months'; and that, when large numbers died at the end of the plague, 'their rearguard was brought by innumerable rats, mostly diseased and emaciated' from 'a form of diphtheria'.

The travelling salesman who had stayed at the Corfield and Fitzmaurice Hotel looked back on the rats as part of a landscape of superabundance, including the flock bronzewing pigeon that had 'obscured the sky, going past from west to east in millions'. When he recalled visiting Winton during the plague, he omitted Ching Foo and shifted the location of the mass killing to W. H. Corfield's store which 'had an earthen floor, and they had dug up a shallow well in the centre which contained water and some contrivance with bait that enticed the rats to tip themselves into the well; and in the morning they wheeled from the store barrows of rats'. A contributor to Townsville's *Daily Bulletin* omitted both Ching Foo and Corfield: 'One day I offered an aboriginal ten shillings if he caught 100 rats before next morning. The following day he came to claim the money and showed me a wheelbarrow full of rats.'

Part II

The Great Irruption

CHAPTER 8

Year One

Each plague was different, but Edward Palmer's paper to the Royal Society of Queensland suggests that the biggest did not happen abruptly. Instead, in a year that was reasonably wet, the rat's numbers grew in western Queensland, though colonists remained largely oblivious. Then huge rains and the prodigious plant growth they engendered the following year spurred the rats to spread far and wide. Much as 1869 saw a modest start to an immense plague, so did 1885, but to one that extended over even more of the continent and lasted far longer.

For three, if not four years, eastern Australia had been drought-stricken. Kenric Harold Bennett was one of many settlers affected, though in unusual fashion, after he managed one more station near Goulburn in New South Wales, applied unsuccessfully to work for the government as a stock inspector and forest ranger in the colony's west, then secured the job that he most desired. In 1883, the Australian Museum hired him on Edward Pierson Ramsay's recommendation to go to the Barrier Ranges to collect 'every department of natural history'—primarily mammals and birds but also freshwater shrimps, earthworms, leaches, spiders, resin of sandalwood and the 'sands of this locality'.

Bennett was upset to be paid just 10 shillings a day, but eager to secure a great haul so the museum would continue employing him. The drought stymied him. A fortnight after leaving Hay, on the Murrumbidgee, he was still on his family's property Yandembah, where he had already collected. A month later he had only reached Moolah, which he had managed for his uncle William Brodribb. As Bennett described it, the country was in 'a wretched state', 'a perfect desert', 'not a blade of grass to be seen'. He never went further because the museum ordered him to return rather than risk his horses and, with Ramsay in Europe and removed from its decision-making, it terminated Bennett's position, leaving him raging that he had been wrongfully dismissed.

A year later, the chief inspector of stock in New South Wales reported that the country between the Darling and the Lachlan was 'in a most frightful state, sheep dying in thousands, and the runs not one quarter stocked'. Many pastoralists in New South Wales and also Queensland were ruined. Some quit their debt-laden properties, which their banks then seized. Most sent their stock coastwards for grass and water. Teamsters stopped plying the roads because it was too dry to travel and there was little market for new supplies. Yandembah was devastated, as an upstream landholder had dammed its one creek, removing all its water, and its well failed. The Bennetts cut their flock from 12,000 to 5000, but only 1200 survived.

Winton was one of many depressed towns, 'just existing on itself', 'as dull as a bottle of soda water that has been uncorked all night', because of the dearth of passing traffic. Insofar as there was water, it tasted 'something beastly', partly because of the number of dead animals in Winton's waterholes. By December 1884, Chinese gardeners, who were the prime providers of fresh fruit and vegetables in Winton, as in most outback towns, had been given 'notice to use no more water', putting an end to production. It was also 'difficult to procure decent meat', flour was 'very scarce', kerosene 'nearly out' and none of the stores 'had a drop of grog'.

Scientists, using old weather records to study climate history, have characterised 1884 as a drought year in Australia's south-east caused by an El Niño. They have identified 1885 as another El Niño year—the driest of the drought years of the first half of the 1880s, involving increased maximum temperatures and lower minimum temperatures. Consistent with this characterisation, there was much talk of drought in western Queensland in 1885. Yet the year began and ended with heavy rain. 'The drought has broken' was the cry that January as the summer monsoon extended into south-western Queensland, northern South Australia and north-western New South Wales.

Alice Springs received four inches of rain, Innamincka on the Cooper seven, Wilcannia on the Darling eight, and Birdsville on the Diamantina fourteen. While corrugated-iron structures survived these deluges relatively intact, many other buildings were blown away or collapsed. Just one chimney was left standing in Birdsville. On the road between Wilcannia and Milparinka, the depression that colonists had called Dry Lake—expecting it to remain that way—was inundated, submerging a hotel erected in its middle, prompting the hotel's licensee to abandon it and rebuild on higher ground beyond the lake's rim. Most Chinese gardeners lost their livelihoods since their plots were on the alluvial flats of rivers. The teams of Afghan cameleers, who provided one of the prime forms of transport through the interior, lost their camels when they got bogged and died. Dams were destroyed. A few colonists drowned. Still, there was near 'universal rejoicing' at the prospect of enduring water and grass.

Winton was one beneficiary. After three inches fell, stations nearby were 'A1 for grass' with the country 'looking magnificent'. The town was surrounded by 'emerald', reminiscent 'of the green fields of the old country', if only 'at a distance'. The local economy was slower to recover. At first, business remained 'very dull', with 'no stock moving'. West of Winton, men walked the plains seeking work, humping their swags twenty-five to thirty miles a day to find that pastoralists would hire no one until their sheep returned. Then traffic picked up, and Winton expected to 'be once more the foremost city' of western Queensland.

The colonists' optimism continued into spring. 'We are going to enjoy one of the rare good seasons,' wrote a journalist in Blackall on the Barcoo, as more heavy rain fell in March. With grass and water abundant that May on the Georgina and along the lower Diamantina, and the Cooper 'in a chronic state of flood for many months', 'fat stock' were everywhere. Then grass and water became scarce in many places but not all. One report came from Hergott Springs, named by the explorer John McDowell Stuart in 1859 after another German artist-scientist, Joseph Hergott, then renamed Marree, after its local Aboriginal name 'Marina', because of anti-German sentiment during World War I. After travelling to Birdsville in Queensland and back to Hergott Springs, a South Australian surveyor reported that 'the country between was looking very well', 'the whole of it splendidly grassed', with 'plenty of herbage'. The Warburton River 'was flooded, and all the waterholes in the intermediate country were full'. The mining town of Milparinka in north-western New South Wales received more than eighteen inches of rain for the year, three times its average.

The one contemporary report of a new irruption came in August from Winton, though the rats were on pastoral stations some distance away. The 'advanced guard' had reached Brighton Downs on the headwaters of the Diamantina, and they were 'present in millions' at Diamantina Gates further downstream. The local correspondent of Rockhampton's *Capricornian* elaborated: 'Their numbers are legion, dogs and cats get so satiated with the slaughter that after a time they allow the vermin to walk over them with impunity.' He considered the rats 'disgusting little animals'.

Later accounts reveal that they also abounded elsewhere in western Queensland. A government party encountered 'myriads of tracks of migratory rats' after crossing the Burke River below Noranside in August 1885. The rats were 'worrying our food supplies and our sleeping hours over a space of upwards of 100 miles', one surveyor wrote. 'The softer parts of the soil were literally honeycombed with their burrows, and it was anything but pleasant to ride a horse across the rat-warrens. The direction of their migration appeared to be from north-west to south-east but this feature, no doubt, was clouded by the course of the rivers.'

Another surveyor recorded that, at the end of 1885, the rats were 'in battalions' around the Paravituary Waterhole on the Georgina River. Because the rats were 'just as numerous fifteen or twenty miles from any water as they were on the banks of the Paravituary', this surveyor concluded that the rats did not need water, when they must have found an alternative supply.

Their 'insatiable appetites' saw them eat 'rations, saddlery, boots, waterbags'. A boy caught dozens each evening as he 'used to amuse himself after supper by propping up a box with a stick attached to a piece of string within the light of the camp fire, and dragging the stick suddenly away whenever rats ventured underneath the box, tempted by the bait placed for them'.

The rats also spread in South Australia along Eleanor Creek, an overflow of the Diamantina within Clifton Hills station, which received sixteen inches of rain in the deluge of January 1885. A member of a private survey party, defining station boundaries at the end of the year, recorded 'there had been a plague of rats... and in one of the wells we found the decaying bodies of thousands of rodents. Each bucket we hauled up contained half water and half rats. The stench was awful. Even the camels refused to go near the troughs. We worked on it for hours, but the supply of rats did not diminish. They had evidently scented the water, tried to get a drink and had fallen in.'

The explorer David Lindsay also encountered rats around the Finke River in the Northern Territory, where the next deluge of December 1885 caused the Finke to flow into its tributaries instead of 'losing itself in sandhills'. Having been without food for three days, Lindsay and his men 'fell in with a party of natives and had a meal of bush rats', 'a little larger than the ordinary house rat', which Lindsay considered 'very good eating' with flesh 'white and tender' and 'rather rich'. Lindsay described a landscape of great abundance with 'splendid grasses, herbs, and bushes, and fine timber, water in claypans and native wells' and the air alive 'with the music of many birds, magpies, parrots of various sorts and many others'. So lush was this country that, as the men rode their camels to the Gulf of Carpentaria, the vegetation was often above their heads. Lindsay saw a link between the abundance of rats and the wellbeing of Aboriginal people. 'When the rats are numerous', he observed, 'the natives become very fat'.

CHAPTER 9

Year Two

On 20 January 1886, a telegram arrived in Sydney from Wilcannia on the Darling, though its message, carried initially by coach, came from two hundred miles further inland. The sender was the mining warden in Milparinka, who usually filed his reports by post and wrote mainly about the local goldfields. The warden telegrammed that a week before, while at the goldmining town of Tibooburra twenty-five miles further north, he had learned the Bulloo River in Queensland was coming down 'bank high' and 'five miles wide' preceded by 'an immense number of rats'. When he wrote, some of those rats had already reached Milparinka.

This telegram was the start of something new: a mass of rat reportage from diverse places. The proliferation of newspapers was crucial. Even small country towns sustained them, while the main newspapers in the colonial capitals not only published extracts from letters by pastoralists and reports by government officials but also employed travelling rural writers and had resident correspondents in many towns. City newspapers also regularly republished the rural press's most interesting articles so that, even when no copies of these rural newspapers survive, their rat stories often do.

Many of the reports involved the Bulloo River where members of the Burke and Wills expedition had encountered the rats. The Bulloo flooded in late December 1885 after eight inches of rain in four days and, by Christmas, had broken its banks and was two miles wide in places. This flood was 'the heaviest for the last twelve years' if not the river's biggest 'since the white man made its acquaintance'. If the rats set out from the Bulloo at the start of that flood, they reached Milparinka in three and a half weeks, travelling an average of five miles a night—a big distance for small creatures.

More reports came from Thargomindah, a three-hotel, three-store town on the Bulloo, with a courthouse and a bank. Largely built of brick rather than iron, using mud from the Bulloo, and sited on a low ridge so it escaped most of the floods, Thargomindah excited attention in 1881 by staging Australia's first camel race, with a £15 prize. Otherwise, one visitor observed, Thargomindah was 'extremely hot and, like all bush towns, very dull'. On 23 January 1886, the local correspondent of Wilcannia's *Western Grazier* wired: 'Tens of thousands of rats are going your way, travelling below Woodburne and Tickalara, from Cooper Creek.' Yet rats continued to be on the Bulloo 'eating the grass up in places very rapidly' and 'not confined to any particular portion of the district'. According to the local *Herald*, they ate part of the boots of a Thargomindah mailman while he was driving his buggy. 'Latest reports state that the driver is recovering from the attack, and his life is not despaired of,' a Sydney newspaper claimed. 'The rats died.'

Birdsville on the Diamantina River was a three-hotel, two-store place, with no school or church, like most such towns. 'No Minister of the Gospel has ever thought it worth his while to come out here', reported a journalist who made the journey. The rats were said to be 'coming down in thousands' at the end of March, implying they too were coming from the north. As they did 'a deal of mischief', the 'Chinamen's gardens suffered considerably' and it became 'impossible' to sleep when 'camping out'. The mailman responsible for the route from Birdsville to Innamincka in South Australia declared it was 'not safe owing to rats having poisoned the wells', a judgment underlined by the discovery on the track of two Chinese men who had died of thirst. By May, the rats had 'obtained complete mastery over all the cats and dogs in the district'. As often, many colonists thought of the rats not as products of a particularly wet season but as harbingers of one. Others reckoned their appearance inexplicable.

An extended account came that May from an overlander taking cattle to Victoria. Having reached the Channel Country with his stock in splendid condition, he reported: 'The country about the Georgina is perfectly alive with rats, and saddlery and rations have to be constantly watched, otherwise they are sure to be destroyed.' These rats, together with a plague of locusts, had 'destroyed all attempts at gardening', so settlers on the Georgina were having 'to be content with pioneer fare—salt junk and damper'. The country in front of the overlander remained 'in a bad state for want of rain', prompting most pastoralists to send their fat stock away for fear of a drought. A resident of Eyre's Creek, a continuation of the Georgina, reported that rats had 'undermined the whole country', even eating out the roots of the grass, prompting him to wonder whether it would grow again when rain returned.

Itinerant journalists typically suspended aesthetic judgment when describing outback towns. Their readers did not need to be told that they were unplanned and ill-built. But, every so often, writers could not contain themselves. 'It has no scenery. Its buildings are straggling, uncouth, unsightly. Its aspect is sleepy, dull, cold and forbidding,' raged the 'Overlander' of the *Queenslander* about Windorah on Cooper Creek. From at least April until July, there were also 'millions of rats' amidst its two hotels, post and telegraph station and police lock-up erected on rocky ground above the creek's red sand. All the while, an 'intolerable plague' of the rats devastated Queensland's west 'by eating the roots of the grass'.

The settlers' attempts at horticulture remained vulnerable. The rats ruined a vegetable garden maintained by officials at the telegraph station at Powell's Creek—part of the Overland line, 120 miles north of Tennant Creek in the Northern Territory. But another garden at Innamincka on Cooper Creek in South Australia flourished after its Chinese owners fenced it with sheets of corrugated iron, placed end on end, creating an impenetrable wall. Others to survive—or rapidly recover—were at Eleanor Creek where police troopers had a good crop of vegetables growing that December.

The rats spread even as the rains of late 1885 did not continue in early 1886 and talk of drought grew. Then, unusually, it rained across eastern Australia through winter into spring prompting scientists to classify 1886 as a La Niña year. 'Water is everywhere plentiful, grass and herbage springing up with magical rapidity and that too, at the season of the year when growth is usually dormant,' reported the *Queenslander* that September. But far from the rats surging in response to the abundant feed in western Queensland, they diminished. The Diamantina was one of the few areas where they still 'swarmed', though it was 'almost free from them' by year's end.

A hallmark of this irruption was the first encounter between the rat and the rabbit—then considered another rodent. Already in the eighteenth century, the French naturalist George-Louis Leclerc, Comte de Buffon, had warned that, in environments suited to them, rabbits ate out the grass and destroyed herbs and roots, grains and fruits, shrubs and trees, creating deserts. Settlers in Australia either did not know or did not care. They were so keen to practise traditional English 'field sports' that they tried again and again when their first attempts to import and release rabbits failed. After they succeeded, a day's rabbit shooting became 'a favourite amusement' for the colonial gentry for a decade or two. Some landowners valued this entertainment so highly they had the police prosecute anyone caught rabbit shooting without their permission. Then the rabbit got out of control, occupying about two-thirds of the continent within fifty years.

The rabbit rendered some country so unproductive that landholders abandoned stations that had once carried tens of thousands of sheep. Other pastoralists retained their stock but feared that 'King Bunny' would force them off the land. Native animals were also affected. While the rabbit did not compete directly for food with the greater bilby which largely eats insects, colonists soon noticed that the rabbit was displacing it. Already in 1882, a settler back of the Lachlan observed that rabbits occasionally shared the bilbies' burrows but the norm was eviction. In fifty burrows he examined, rabbits were the only occupants. Where the bilby had been found in prodigious numbers just a few years before, it was 'now seldom seen'. Little wonder that, in *They All Ran Wild*, Eric Rolls concluded that the bilbies 'died out because rabbits had taken over their burrows and they had nowhere to rear their young'.

At least one colonist had a different explanation. He thought that sheep did 'not appear to agree with the bilby', prompting it 'to clear off to other country'. Scientists have more recently suggested that cats and foxes also contributed, but there is no evidence of that in accounts of the back country of the Lachlan in the late nineteenth century. Most colonial observers thought the rabbit was to blame for driving the bilby northwards, by taking possession of its burrows and then adding multiple exits. With sympathy on the side of the 'harmless' bilby, 'well-known to all old settlers', a contributor to the *Australian Town and Country Journal* remarked: 'Brer Rabbit has developed Australian traits, he won't burrow himself, if he can get a bilby's hole, out of which he drives the aboriginal inhabitants, just as we drove the blacks.'

In 1883, the New South Wales Government turned to the law, as it had done thirty years before with the dingo. A new Act of Parliament required landowners to destroy rabbits on their land. Because the 'rabbit nuisance' was a collective problem, the legislation tried to spread this cost equitably, obliging all landowners to contribute to a Rabbit Fund whether or not their land was infested. The law also tried to ensure that rabbiters—as rabbit trappers became known—had an incentive to kill as many as possible by stipulating they would be piece-workers, paid for each rabbit they killed. The government created a special inspectorate to ensure the work was done—though why it was located in the government's Mines Department was 'a question frequently asked, and never satisfactorily answered'.

Several of the new inspectors became close observers of the rats, starting thirty miles from Milparinka at Cobham Lake, which the prodigious rains of January 1885 filled, then the rains that December replenished. As the lake in 1886 was eight miles long, one and a half miles wide and thirty feet deep in places, colonists expected the water to last at least four years. But the surrounding country was soon so dry they considered it drought-stricken. In early April, Rabbit Inspector Holding reported the rats were around the lake 'in great numbers'. He identified them as 'Queensland rats (Bulloo River)', like the mining warden at Milparinka at the start of the year. At least a few colonists, for once, ate the rats out of curiosity rather than need. A Sydney newspaper heard they were 'delicious eating—far better than rabbit or kangaroo'.

Despite the classification of 1886 as a La Niña year, the dearth of rain was widespread that autumn. The plains along the Darling River were 'one vast desert with not a green blade to be seen for hundreds of miles'. The river itself was a chain of waterholes. So too was the Lachlan River, which was the lowest colonists had ever seen. The Murrumbidgee was its driest for twenty years. There were dust storms from Wilcannia to Hay, prompting one settler to write: 'We have breathed dust, heard dust, eaten dust…We have had dust hash, dust pie, jugged dust, dust pudding, dust-a-la-mode, dust omelette, quail on dust and buttered dust.' The journalist, who wrote a twelve-part report about this 'Drought Land' for Sydney's *Daily Telegraph*, first heard of 'the invasion of the rats from the Queensland border' at Cobham Lake, where the rats had 'played havoc with the Chinese garden'. He observed their tracks everywhere across the thirty miles to Milparinka and learned that, while the rats hid by day in the saltbush, they travelled all night, heading south towards the Darling River.

Silverton in the far north-west of New South Wales was the colony's boom mining town—the site of a quarter of the mining leases granted in New South Wales and, just two years after its establishment, nine hotels, three breweries and a small theatre. While the surrounding country was 'noted for its sterility', it was transformed by unusually heavy winter rains, which put an end to talk of drought in New South Wales. 'A wet winter!' a journalist on the Darling River exclaimed. 'An inch of rain at the end of July or the beginning of August is worth five at other seasons; it gives us good grass in the spring,' observed another on the Murrumbidgee. These rains, which 'literally flooded the country' around Silverton, resulted not only in 'an abundance of feed' but also a fabulous display of native orchids, heath and Sturt desert pea, creating a 'paradise'.

When a few rats appeared in early July and many more arrived in August, the local *Silver Age* considered their origins and destination unfathomable—a typical piece of rat writing, but also an acknowledgment of the colonists' limited capacity to track the rats. 'Where they come from nobody seems to know, and their disappearance is equally mysterious and rapid. A large tract of country may be overrun by them one day, and in a week perhaps not even a trace of them is to be found.' They remained a staple of Aborigines who ate 'a great number, which they highly prized for their flesh'.

Terriers were in demand in Silverton to counter this 'great North-West rat army', while rat-catching replaced horseracing as the miners' most 'popular pastime' and 'only fun'. There were so many rats that 'a score or two' were 'taken morning after morning from some shallow prospecting shaft or pot hole, into which they… strayed during the night', and they were 'almost to be kicked from every bush affording a sufficient cover'. The *Silver Age* recorded they were 'of carnivorous habits, even preying with avidity on their dead or wounded comrades'.

'Immense numbers' reached just north of Hergott Springs in South Australia that January. As they often did, commentators expected floods to follow, 'a happy result' for pastoralists in the far north. Early in February, one of these pastoralists advised that an 'army of rats' was going through his station travelling south. 'The same kind...came down in 1870,' he recorded. He thought they had 'brought a run of good seasons' and hoped they would again. 'The cats are having a gay time with them,' he added.

The rats also returned to Parallana station, which had received twice its usual rainfall in 1885, including a ten-inch fall in four days at the start of the year. 'About fourteen years ago we had a similar visitation, nothing at that time was secure from their depredations—they undermined the store, and every building was infested,' wrote a local resident in March 1886, looking back to the plague of the early 1870s. 'This time they are not, as yet, quite so troublesome, probably owing to the fact of there being more feed about.' To him, the rats were also a mystery: 'I do not think anyone knows where they come from. They come in vast armies, leaving their tracks for hundreds of miles.'

A resident of Innamincka, who travelled to the town of Farina that April, confirmed the rats were travelling south. 'Hundreds of thousands' had reached Hergott Springs, which had become the terminus of the Great Northern Railway from Port Augusta in 1883 and then a hub for further construction. Another report in April 1886 from the southern Flinders Ranges was of 'a kind of plague'. The rats had 'taken possession of everything' from the settlers' coats to their horses' bridles, polluting tanks and wells with their dead carcasses.

Even at the best of times, the conditions of the navvies

constructing the Great Northern Railway were notoriously bad—intense heat, terrible flies, saline water and meagre food. The rats were a frightful addition. That May, the line's chief engineer, J. Randell Mann, reported the rats had eaten fifteen tons of chaff and 100 bags of bran and oats in three months. Mann's attempts to thwart them by building raised platforms with overhanging edges had failed. 'They seem to eat anything that it is possible for a tooth to go through,' he declared. When an eleven-truck train hit cattle that had strayed onto the track beyond Hergott Springs in mid-April and seven trucks were thrown off killing five men, Mann had two men watch over the dead through the night to stop the rats eating their mangled remains.

Another account came from the Camel-carrying Company, which provided cartage through the inland. When its manager reached Adelaide in June 1886 after a seven-month trip through western Queensland and central Australia, he reported that much of the country was undermined by rats creating new burrows as they travelled. The manager elaborated: 'The rats appear to have migrated from the North of Queensland, having been driven southward by the floods. They are now eating the bush, and are a terrible nuisance to the squatters...The drought now appears to be killing the pest off rapidly. Ordinary means of destruction were quite useless, dogs and poison making no perceptible difference in the vast number of the rodents.'

One of the few rats at Strangways Springs on the Overland Telegraph arrived in a teamster's dray in June. A journalist identified it as one of the 'Queensland rats', accepting the commonplace view that they had travelled hundreds of miles. He also reported that the Arabana people called the rats *myroo*. When a local pastoralist was asked by a government inquiry whether he had ever tried to grow wheat, he responded: 'Yes, and the rats ate it all. We sowed the wheat in drills, and the rats began at one end of the drill and ate it all out to the other end.'

Through the rest of 1886, there were almost no reports of the rats in South Australia, suggesting their numbers had plummeted. A rare account from Teetulpa, the biggest South Australian goldfield, which attracted at least 5000 miners to the Flinders Ranges, was of 'field rats that nibble your toes and your boots'. Another account, in December from the colony's far north-east, identified Eleanor Creek, once 'swarming', as 'almost free'.

The prime attempt at describing the rats' spread came in 1918 from John Bagot, an engineer on the Great Northern Railway, who thought his recollections were of 1887, but they fit accounts of 1886. Bagot recalled the rats arriving 'from the north, from the great dried-up river beds of the Finke, the Alberga, and the Macumba, and smaller rivers which, in times of extraordinary rain, pour their floods into the north-west area of Lake Eyre'. Their appearance on the lake's southern shore was so sudden that £1000 of provisions, tents and other commodities were destroyed 'before precautions could be taken'. Before long, they had headed further south, keeping close to artesian springs and temporary surface waters.

According to another account, many colonists feared that 'country towns and the city of Adelaide and all South Australian civilisation generally were in for a reign of terror', but the rats 'never reached the seacoast'. John Bagot recalled that they continued south from Hergott Springs through better country, following the plain between Lake Torrens and the Flinders Ranges on a front up to thirty miles wide until they almost reached Spencer Gulf.

Pellets of owls provide the most striking evidence of the rat's reach. These pellets come from Weekes Cave, a sinkhole in the Nullarbor Plain rich in fossilised birds, about ten kilometres from the coast and 100 kilometres from South Australia's border with Western Australia. Bones of the long-haired rat found within pellets in the cave are a sign of the rat extending to the Great Australian Bight. They are also the oldest sign of the westerly reach of the rat. Since these bones were identified in the early 1970s as about a hundred years old, they could be from the irruption of the early 1870s or from that in 1886.

CHAPTER 10

Year Three

1887 started with more good rain. Thargomindah and Cooper Creek each received one and a half inches that January, and Bulloo Downs two and a half. But these falls were eclipsed by a cyclone that struck the Gulf of Carpentaria in March, killing seven people and wrecking all buildings in Burketown except the customs and court houses, as the Albert River rose ten feet and a tidal wave surged thirty miles inland. When this water eventually receded, settlers found dead fish all over the surrounding plain, but could not measure how much rain had fallen, since Burketown's one gauge had been swept away: a common occurrence in big storms, compromising colonial meteorological records.

As the cyclone reached further inland, its impact was measured and reported with wonder. Monkira station on the Diamantina River in Queensland received ten inches on the seventh and eighth of March, two days after Burketown was devastated. Kyabra station had eighteen inches in ten days. By the end of the month, Winton had more than seven inches, its biggest falls for March since colonists began keeping records. Boulia had a record fourteen inches, Windorah a record fifteen inches. Milparinka in New South Wales received six inches, Wilcannia eight inches in a fortnight, Booligal on the Lachlan River three inches in eight hours.

The mail contractor responsible for Birdsville in south-western Queensland soon reported that the country further south was so heavily flooded that, instead of riding his horse, he went by boat for ninety miles, often rowing across country rather than following the usual course of the Diamantina River. Another mailman, responsible for the far north of South Australia, reported that at least 48,000 square miles were inundated. While one local resident disputed these accounts, claiming the track to Birdsville remained open to bullock and camel teams, another had no doubt that Lake Eyre—already famous for being dry—temporarily was not. A commentator observed that, if a European explorer had gone inland for the first time, he would have 'come back a convert to the ancient theory that the centre of the continent is occupied by a vast fresh water lake'.

The conventional yardstick for such water was other very wet years. The problem was that there were no precise records and European memory was short. The big floods of 1864 were prehistory for most settlers in 1887. The floods of 1870 were recalled by many more. It was known as 'a year to be remembered for feed and water'. The floods of 1887 were judged even bigger. Colonists in eastern Australia identified 1887 as the wettest year in almost a century of European settlement. Now it is recognised as a La Niña year.

Some colonists grew sated, especially in New South Wales. A journalist in Balranald on the Murrumbidgee reported 'complaints of too much water' from inundated landholders that March. 'The Lachlan people have had enough of water this year,' remarked another journalist in December. 'Never since a white man put his foot in the Riverina has such weather prevailed in this province.' Most settlers were delighted. 'Everyone looks forward to a magnificent lambing, and no doubt the number of sheep throughout the colony will be something marvellous,' ran a report from the Murrumbidgee in April. On the plains of north-western Queensland, the seeds of the different forms of Mitchell tussock grass, which colonists reckoned the 'richest natural fodder in the Southern Hemisphere', began springing to life. In central and far western New South Wales, the native grasses reached knee-height, waist-height, then shoulder-height. Having enjoyed a boom year in 1886, which saw cattle likened to 'baby elephants', pastoralists stocked the land more heavily than ever. Still, their sheep and cattle could not eat a quarter of the feed in some areas and their mutton and beef were 'too fat for many constitutions'.

Reports of the rats in Queensland began again that May when they passed through the township of Aramac, seemingly travelling south, and crossed the flooded Herbert River in hundreds. By August there were 'large numbers' at Birdsville, as there were that October, when they were also 'in millions up on the Flinders, in the Burke district, and down on the Warrego and Paroo in the south'. A pastoralist outside Richmond reported in November that he had killed more than six thousand, yet they seemed 'as numerous as ever'. Within a few weeks, another pastoralist declared them 'equally bad' on the Barcoo where station managers were 'at their wits' end to know how to cope with the pest'. Because a team of drovers on the Cooper could secure no other food, they 'fed on rats for three weeks', a rare instance of colonists turning to them as famine food.

The rats were also accused of a new form of destruction—causing bushfires by nibbling and igniting wax matches, which had all the combustible chemicals in their heads. This explanation of fires was initially advanced in Australian towns and cities in relation to English and Norway rats, then was extended to the countryside in relation to native species. By one account, most stations in western Queensland responded by abandoning wax matches 'during the rat years 1886 and 1887' and, once they began using safety matches, which had some of the chemicals for ignition in the striking surface, fewer fires occurred.

In South Australia, the rats infested Cockburn, a canvas town on the colony's border with New South Wales, providing 'a bitter experience' for a journalist who slept out there. They were again 'in marvellous abundance' on Eleanor Creek where, prepared 'to eat anything and everything...except sand, cartridge-cases and stirrup-irons', they were the 'greatest enemies' of cabbages, pumpkins, melons and tomatoes in the policemen's garden. Rats also invaded the Teetulpa goldfields in the Flinders Ranges, where a team of miners tamed a goanna so 'not a rat came near' and the miners 'had more than one offer of £5 for the reptile'.

When the Victorian Government geologist Reginald Murray visited Teetulpa that January, he heard that 'a perfect army of the rodents recently crossed the country, where from or whither bound no man knoweth'. He was also told that males and females kept 'strictly separate' while travelling, the first account of the gender of the rats. Because they were not marsupials, Murray thought them 'evidently of imported breed', most likely the 'common house-rat'.

But other colonists remembered them as a native species, which 'lived in warrens', came from 'the north, the centre of Australia' or 'the Diamantina country, South-West Queensland' and, after reaching Teetulpa, continued travelling south but 'did not get far' before they 'mysteriously melted away'.

The rats excited most attention in New South Wales where the first report, dated 3 January, was from Menindee, where the government had just embarked on its first attempt at damming a significant source of the Darling River. After giving a hundred unemployed men free rail passes to Bourke, and then free steamer tickets down the Darling, it set them to work at the Menindee Lakes. A royal commission into conserving water in New South Wales had just proposed the erection of sluice gates on the lakes, but the government tried to hold back the water with sandbags. When they failed almost immediately to much derision, the government built a solid barrier.

Ludwig Becker, who was in Menindee over the summer of 1860–1861 with the Burke and Wills expedition, recorded 'one hotel, one store, a kitchen, and two native huts'—a rare inclusion of Aboriginal people in such enumerations. Twenty-six years later, Menindee's itemisers listed two hotels, two stores, a bank, a post and telegraph office, Mechanics' Institute, police barracks and school, and the rat. The local correspondent of the *Riverine Recorder* reported: 'We are suffering from a plague of rats which have invaded the place and are the terror and annoyance of all housekeepers.' With the *Recorder*'s readership in mind, in Balranald on the Murrumbidgee River, he predicted: 'As they are travelling southwards, I dare say you will receive a visit from them in time.'

The rats were 'still to the fore' in Menindee at the start of February. But small numbers also reached Ivanhoe, a horse-changing station for Cobb and Co between the Darling and Lachlan rivers. By mid-March, large numbers were upstream of Menindee, on the Darling around Bourke, the hub of the wool industry in inland New South Wales, and the first town of consequence the rats had approached. They were 'to be found under almost every piece of fallen timber', but did not 'worry round the houses much', and by the start of April had 'swept by', leaving one local journalist wondering what would happen when they continued 'their march to civilised parts'.

Hay on the Murrumbidgee, more than 350 miles south of Bourke, was a test. 'One of the best laid out and smartest looking places in New South Wales', exceptional for its 'broad thoroughfares lined with shady trees', Hay thought of itself as a cathedral or a pro-cathedral city—the centre of the Anglican diocese of the Riverina. It was also the site of Presbyterian, Wesleyan and Roman Catholic churches, a convent school with two hundred pupils, and had Temperance and Masonic Halls, an Athenaeum and Public Library with over a thousand volumes, a public park and a town council and it was illuminated by a local gas company. 'We never heard of rats until a few weeks ago,' a surprised resident wrote at the start of April. The *Riverine Grazier* wondered whether there were 'fanciers enough to start a rat pit' as a new form of local entertainment. 'Cats, at last, have some employment,' the *Grazier* observed a week later of Hay. 'The Bulloo rat gives her plenty to do.'

To the west, as predicted, the rats reached Balranald, a seven-hotel, six-store, two-church and gospel hall town, with 'the usual complementary aggregation of more or less insignificant little shops and dwelling-houses'. After arriving in April, the rats remained 'numerous' in early May, 'making havoc with the roots and vegetables, of some of the gardens'. In September, they were again 'quite a plague', and 'terrible' by early October. On one station, they 'completely undermined' a large haystack, resulting in its collapse. On another station, the grass appeared in several places 'like a newly-mown field'.

The sole report of a plague east of Bourke came in May from Walgett—another seven-hotel town where, to the delight of its Progress Committee, the main street was about to be kerbed and guttered, a soap-and-candle factory was set to open and this domain of 'beef and damper' was acquiring its first coffee palace and oyster saloon, boasting improbably 'fresh' oysters at one shilling per plate. The rats 'scourged' the local market gardeners—including a Frenchman, who 'suffered ruinously' as the rats 'devoured a splendid crop of melons and everything else they could reach'.

The agricultural reporter of the *Sydney Mail* attempted to map the rats' course through New South Wales. He wrote, seemingly looking back to January 1886, that the plague was first noticed near Milparinka where it 'took the form of myriads of rodents, all making their way in a south-easterly direction'. At Menindee, at the start of 1887, 'the main body of the invaders divided'. Many went south down the Darling beyond the township of Pooncarie, possibly extending as far as the Darling's junction with the Murray at Wentworth. The remainder headed east into the back country, to places such as Ivanhoe. By June, they had reached Tupra Landing, on the Murrumbidgee.

Another account of the rats' course came from Kenric Harold Bennett, who, after being more out of work than in since his uncle sold Moolah in 1879, was employed by the government as a rabbit inspector at the start of 1884. At first, Bennett was responsible for an area centred on Ivanhoe. Then he took charge of the adjoining district centred on Mossgiel, before returning to Ivanhoe. One of his tasks was to check that landholders fulfilled their obligations to control rabbits and counted those killed before authorising the trappers' payments. Another was to lead a government party that killed rabbits at the expense of landholders who failed to do so. Time and again, he fumed at the futility of most of his rabbit work and lamented how it curtailed his opportunities to observe and collect, while still doing so whenever he could.

One of his interests was directly related to his work: the wedge-tailed eagle as a control of the rabbit, a deeply contentious subject. In an era when colonial legislation protected only a tiny number of native birds and mammals during their breeding seasons, some settlers argued that the eagles warranted protection as one of the pastoralists' 'best friends'. Others argued the wedgetails should remain unprotected because they were a threat to lambs. Bennett informed this debate, reporting that an eagle's nest he examined contained the skulls of one lamb and twenty-one rabbits.

Bennett also continued collecting for the Australian Museum—always for payment, having abandoned his initial gentlemanly stance of buying his own equipment and donating what he found. As Edward Pierson Ramsay peppered him with requests, Bennett provided fungi, butterflies, moths, birds and eggs—once climbing 'the face of an almost perpendicular rock' about seventy feet to reach the nest of a peregrine falcon. With Ramsay as his intermediary if not agent, Bennett also found a market both in Australia and in England for bird skins and eggs that Ramsay did not want for the museum or his private collection. He assembled by far his biggest botanical collection comprising 191 specimens and, responding to criticism of his contributions to Sydney's 1879 International Exhibition, made extended notes about where each one grew, its suitability as fodder and its response to drought. Then, Frederick Turner of the New South Wales Department of Agriculture named each specimen before the Australian Museum implicitly acknowledged the significance of this collection by displaying it at the Colonial and Indian Exhibition in London in 1886. Bennett sent other plants to J.H. Maiden, the curator of Sydney's Technological Museum, who had become his conduit to the Linnean Society when he wrote about botanical matters. But the impacts of colonisation meant Bennett could secure few small marsupials, the specimens Ramsay particularly wanted. They had become rare, Bennett explained, due to competition from the rabbit and legislation which identified them as 'noxious' and offered bounties for their scalps, since they competed for pasture with the settlers' stock.

Bennett also collected Aboriginal tools and weapons—most likely buying some at minimal cost, bartering for others, and finding others abandoned. Ramsay was particularly eager for this material because a fire in 1882 had destroyed the Australian Museum's ethnological collection. Bennett took particular pride in a stone chisel in a wooden handle which was still 'in use by an old man' when he obtained it. He found other objects including big grinding stones which Aboriginal women often carried great distances, then 'left at one of their favourite camping grounds, to be in readiness for the ensuing seed season'. While Ramsay sought only the pre-European, Bennett was also interested in the adaptability of Aboriginal culture—collecting a net made of colonial wool, which Aborigines obtained by unraveling discarded European clothing. In total, Bennett acquired more than 175 weapons, tools, ornaments and toys used by men, women and children between the Lachlan and the Darling. While most were wooden, many were stone, and a few were made of skins, bone, fibre and yarn. When he looked to sell this material to the museum, he typically had no idea of its market value and relied for a fair price on Ramsay, who acknowledged the collection's significance by publishing a small catalogue of it in 1887.

Bennett also paid close attention to the rats. He explained that 'for some months previous to their appearance' in February 1887 in Ivanhoe, he had 'heard of their advance in a southerly direction from Western Queensland'. When the rats began arriving there, they were just 'scattered individuals' but many followed. Although Willandra Creek, an anabranch of the Lachlan, was in high flood, it did not stop them. In mid-April, the country west of the main road from Booligal to Wilcannia was swarming. As usual, they travelled by night while taking refuge during the day in rabbit burrows and cracks and holes in the ground. By mid-May, 'the main body had passed on…in full march for the Lachlan', though 'numerous stragglers still remained'.

Bennett himself had also moved, following complaints that his pursuit of natural history led him to neglect his job as a rabbit inspector. While the local *Riverine Grazier* had lauded Bennett's rabbit work in 1885 and the *Hay Standard* did so again in 1887, his superiors disciplined him by sending him to Tilpa on the Darling River between Wilcannia and Bourke, making him inspector of an even bigger, worse affected district. A local resident, eager to boost Tilpa, identified it as blessed with 'a good store, a telegraph office and a commodious hotel'. Bennett initially described Tilpa as a 'one horse place', with just one 'tumbledown' hotel, 'not by any means as civilised' as Ivanhoe, while its surrounding country was 'far more dreary and uninteresting'. When he knew Tilpa better, he judged it a 'god-forsaken hole'.

J. J. Coker, Tilpa's rabbit inspector since 1884, was meant to be a beneficiary of Bennett's punishment. He moved to Ivanhoe which was triple Tilpa's size with two hotels, two stores, a telegraph office, police barracks, blacksmith and several houses. But this 'civilisation' was no reward for Coker who, like Bennett, was aged about fifty, but was in bad health, suffered from depression, and was subject to fits. Unable to find a house, he began by sharing a room with the local sheriff's officer. One night, soon after his arrival, Coker attempted suicide, cutting his throat, 'not dangerously', while the officer remained asleep in the same room.

Bennett, more than ever, detested the 'infernal rabbit business', derided it as pointless, loathed and lampooned his superiors, and vainly implored Ramsay to offer him alternative employment. Yet he still coped better than Coker—sustained by his passion for collecting and observing, the 'only pleasures' in his 'otherwise dreary life'. The rats particularly interested him. In May 1887, they were 'tolerably numerous' along the Darling at Tilpa and 'for some short distance out', but seemed to be 'almost unknown' in the back country towards Cobar. While the Darling was 'in high flood, and the water extended out for miles,...this did not stop the onward march for they soon appeared on the opposite side'. In mid-July, they were 'much more numerous' along the Darling, and 'spread further out' and Bennett wondered whether this was 'another invasion taking a more easterly direction than the preceding one'. At the end of September, the rats were still at Tilpa but 'not nearly so numerous'.

Just as they had on South Australia's Western Plains in 1871, colonists evoked the rats' prodigious numbers by describing how they eradicated all traces of what had gone before. Bennett provided an example from Kilfera—a 1300-square-mile station just south of Ivanhoe, carrying 140,000 sheep, which had been within his domain as a rabbit inspector until he was banished to Tilpa. Bennett recorded that one day a mob of sheep was 'put through a gate near the house...and of course thousands of tracks or foot-prints of sheep were visible on the dry dusty soil through and around the gate; but the next morning not a track was to be seen', because of 'the swarms of rats that had passed during the night, millions of tiny foot-prints completely smoothing the dusty soil'.

As in 1886, some colonists ate the rats out of curiosity, then for pleasure. In the west of New South Wales, men 'from the back blocks' reckoned the rats 'not at all bad eating, possessing the flavour somewhat of a rabbit or the ordinary kangaroo rats of the bush'. Just as Bennett's account of Kilfera provided further corroboration of how the rats would have erased all marks of the return of Burke and Wills to their depot on the Cooper in 1861, here was more confirmation that, had the explorers eaten more rats, they might have enjoyed their novel diet and survived.

CHAPTER 11

Year Four

Pastoralists entered the centenary of British Australia in 1888 with confidence, it had been so wet. 'Old hands' had seldom seen the country 'so green in the middle of summer'. Even without more rain, there would be 'lots of the best feed' for a few months, enabling cattle and sheep to be fattened for market in what were normally the leanest months of the year. More rain in many places sustained this optimism that February. Then 1888 became eastern Australia's driest year as measured by the colonists' meteorological records, prompting its modern characterisation as 'The Centennial Drought'. While the average rainfall of Cowarie, east of Lake Eyre, is 120 millimetres, in 1888 it received sixteen.

Charles Todd, South Australia's postmaster-general and superintendent of telegraphs, as well as its astronomer and meteorologist, was among a small group of colonists seeking to identify patterns and connections in Australia's variable climate. At the end of 1888, he recognised it as one of many years in which a severe drought occurred in Australia while one was occurring in India. Like so many other significant observations of the Australian environment, Todd reported this finding not in a scientific periodical but in an article in a newspaper, the Melbourne *Argus*. The synchronicity he observed between Australia's and India's droughts has been identified as part of the phenomenon now known as El Niño, and 1888 has been confirmed as an El Niño year.

Kenric Harold Bennett was back on Yandembah, his position as an inspector come to an end in the wake of the government accepting its attempts at rabbit control had failed and dismissing all its staff. Like the journalist-drover who visited a decade before, Bennett may have found his family's homestead an embodiment of civilisation, especially after 'one-horse' Tilpa. But surrounding it was devastation. 'I have been driven nearly mad (and in fact have been quite sometimes) in consequence of the terrible drought prevailing here,' he wrote to Edward Pierson Ramsay that October. With only three inches of rain all year, the country for more than two hundred miles to the Darling was in 'a dreadful state'. Having taken over managing Yandembah from his younger brother Edward, Bennett had sold many of its sheep, securing almost nothing for them. He had expected the rest would survive until rain came, but they were dying fast. 'Hope has gone,' wrote Bennett who was trying to support his mother, sisters, wife Annie and their two children.

As he wondered what to do, Bennett imagined emulating Price Fletcher of the *Queenslander* by writing about natural history for the weekly *Sydney Mail*, which had such a column earlier in the decade. He wanted to draw on his own writing, but his 'wandering, unsettled life' meant he did not have copies of his articles published by the Linnean Society, nor duplicates of the mass of material he had sent to scientists in Sydney. When he asked for copies, Ramsay did not respond, despite having encouraged Bennett to try journalism. He also did not reply to another request. Ramsay, it turned out, was ill, so Bennett had 'nothing but memory to rely on, a faculty...fading fast'.

The last months of the 1888 drought were worst. 'A perfect demon of a season,' Bennett judged, 'more disastrous in its effects than anything we have had in that line previously and goodness knows I have seen bad ones enough.' When Bennett mustered Yandembah's flock, he found 5000 sheep, down from 14,000, a 'terrible deficiency'. Birds and mammals had 'fled the scene' since 'it was no use coming to a veritable desert', diminishing opportunities for collecting, but Bennett remained eager to write. One of his longstanding interests was how birds responded to European settlement. After he made extensive notes, lost them as part of his itinerant life, then for once rediscovered them, he asked Ramsay: 'Perhaps they would be of interest to you?' They were, prompting him to embark on his most ambitious paper about environmental change.

The rats remained in large numbers in south-western Queensland, where the usual big summer rains kept most rivers and creeks running early in 1888 and resulted in floods on the Diamantina and the Cooper. In March, a customs officer stationed at Beetoota to examine goods from South Australia as part of intercolonial tariff regimes reported that after more than six months of the surrounding country being 'remarkably free of rats', they had returned 'in millions'. While 'good feed' had 'kept them from molesting people', that feed was disappearing and the rats had begun 'entering houses and helping themselves to everything from flour to water'. They were there for three months, 'all but disappearing' only in June.

Another report came from Barcaldine, east of Winton, which was attracting most attention for its new artesian bore, pronounced 'the best in Australia, and the world', yielding 165,000 gallons a day. By one account, the water, which spurted out of the ground at a temperature of 101 degrees Fahrenheit, was 'perfectly clear and soft, and very slightly mineralised'; by another account, it tasted like 'a weak solution of epsom salts' and was 'generally disliked for drinking purposes'. Either way, it was 'a godsend' used 'by households for all purposes'. A canvas screen was erected around it so residents and visitors could bathe there. Because colonists looked on it as an infinite resource, the water was allowed to run, day and night, whether used or not.

Barcaldine's *Western Champion* was oblivious to local history when the rats appeared nearby. The *Champion* had forgotten that in 1887 its correspondent in Aramac, forty miles north of Barcaldine, had reported that the rats appeared to be heading south and Barcaldine might be invaded soon. It did not realise that the rats had been to the south on the Barcoo in 1868 and again in 1887. It thought of the rats as a blight far away. The *Champion* considered it 'well known' the 'plague of rats' was 'confined to the country to the west of Winton, and very little to the south of that area'. That 'an army' should have 'determined upon incursions in the direction of the Barcoo country', south of Barcaldine, was 'strange and unprecedented'.

The *Champion* reported that 'hundreds of thousands of the rats passed across the desert country' between the Alice River and

Patrick Creek, then followed the Blackall Road in such numbers that the road looked as if 'a huge flock of Lilliputian sheep had been passing along it'. In the midst of this barren country, there were hundreds of dead rats, 'perished probably from thirst in the long dry stage'. Another journalist thought the rats were travelling west towards the 'dry desert country'. Because he regarded it as 'well known that the rodents cannot live long in an arid district', he concluded that the rats were heading there because they could sense rain would fall.

The rats were also in large numbers in the Gulf Country, thought to be arriving from the Northern Territory. The news came from Camooweal—a Queensland town notable for its 'backward move' in 1886 when one of its two hotels closed, gifting the other a 'roaring trade' until it ran out of beer and weeks went by before it secured new supplies. At the height of this plague, which continued from May until August in Camooweal, a man killed one hundred in an hour with three twisted strands of wire. A stockman would recall enjoying 'a sumptuous repast—a big dish of rats with...a rich heavy thickening of flour', while rats were 'journeying across the Barkly Tableland in armies'.

Irruptions also continued in New South Wales. At the start of March, another plague was 'fast approaching' Coonamble on the central-western plains, 'advancing like the rabbit, only from the opposite direction'. Rats soon also appeared again in Walgett. They were also in the Riverina around Argoon, Tamora, Moulamein and Deniliquin where they were initially thought to be English or Norway rats, 'imported...from the seaboard in packages of merchandise', but then were identified as part of 'the horde of vermin which infested the Upper Darling and Paroo country during the last good season'.

Victoria also experienced a rat plague. To get there from south-western New South Wales, the rats had to cross the Murray River. In late 1887, they were close to its north bank. In early 1888, they were on the southern side, and had taken 'a firm hold' on Benjeroop, especially on farms with river frontage. They attracted most attention in April and May when they were around Wharapilla, Gunbower, Terrick Terrick, Murrabit West, Swan Hill, Cohuna and Kerang. In 'hundreds' by one account, 'marching in thousands' by another, and in 'millions' by a third, the rats were 'devastating the country', 'doing great damage to grain and hay', with burrows 'all over the cultivation paddocks' and also infesting houses so one farmer had 'his dogs tied up at the doors of his house every night to keep them away'.

As usual, some colonists thought the rats' origins a mystery. Others thought they had been approaching for months from New South Wales, 'crossed the Murray somewhere in the Swan Hill district', and before that 'come from the direction of the Darling'. They were usually described as 'Queensland rats', though the *Kerang Observer* thought no one was 'clear as to how they came by that name'. While said to be travelling further into Victoria, there were no reports of them doing so. Even so, if they were long-haired rats, they had reached further south then ever recorded before or since.

CHAPTER 12

The Name of the Beast

Humphry Davy—not the British inventor of the miners' safety lamp and discoverer of laughing gas but a British namesake who came to Australia as a five-year-old and went on to manage several sheep stations—enjoyed some renown of his own in the 1880s. A regular correspondent of local and metropolitan newspapers, eager to declaim on many subjects, Davy compiled Aboriginal word lists, reported on Aboriginal traditions, called for greater government provision for Aboriginal people, advocated irrigation and wrote at length about pastoral matters. He attracted most attention in 1885 when he investigated the approach of the rabbit from New South Wales for the Queensland Government. His report was acclaimed for prompting Queensland to enact rabbit legislation and build a rabbit-proof fence on its border. It was pilloried for advising the rabbits were still 130 miles away when, within a month, a New South Wales inspector found that country fifty miles away was infested.

Davy drew on this fieldwork in April 1887, when he wrote from his home near Balranald to Hay's *Riverine Grazier* about the 'mysterious rats' that had just reached the Murrumbidgee, and sought to put an end to their mystery. He declared: 'These rats belong to Queensland, and are very largely habitants of an area of open low-lying country...about 100 miles west of Hungerford.' More precisely, he claimed the rats came from the Bulloo, which was often just a series of waterholes with no flow between them and, when flooded, did not flow into another river but spread out over a claypan, forming a vast, temporary, shallow lake up to one hundred miles long and seventy miles wide. This area, which extended from Queensland into New South Wales, was then known as the Carrypunday Swamp. Now the more northern part is generally known as Lake Bulloo, while the more southerly part in New South Wales is called the Bulloo Overflow.

The Karenggapa people had shaped this landscape by building a dam wall that could hold 150,000 gallons and last many months despite evaporation and seepage. The Mardigan people had built stone walls, weirs and pens from river rock to channel yellowbelly, perch and catfish into areas where they could be easily caught. Yet most of the Carrypunday Swamp was subject to natural forces and it was so big and shallow that wind, evaporation and absorption soon dried it. When William Wills passed by in November 1860 as he headed for Cooper Creek with Robert O'Hara Burke, much of the swamp was filled, but Wills recognised that this water 'could not last throughout a dry season'. He also observed that the rich vegetation around the swamp sheltered an abundance of fauna including 'numerous kinds of waterfowl and snakes'. By March

1861, when Hermann Beckler followed in the rearguard of the Burke and Wills expedition, the Carrypunday was 'an extensive dry marsh', though its banks remained thick with plants, 'creating a marvellous expanse of green'.

Humphry Davy identified the swamp as 'the abode of the rats', having 'carefully inspected' a large part of it in August–September 1885 when it was dry. He reported that, rather than make burrows, the rats constructed nests that could be seen 'in places for many miles round'. Davy described these nests as cone-shaped mounds, made of driftwood and debris. He thought they were 'not unlike a swan's nest, but three times as large and high'. Although Davy did not identify their makers, these structures were clearly the creations of stick-nest rats.

A few years earlier, a storekeeper described being invited to 'tuck-out' by some of the Aboriginal people who continued to live 'in the old-style' on Bulloo Downs. Having declined to eat snake because it looked 'dirty', the storekeeper started with possum and lizard, then finished with rats. Echoing Parisians during the siege a decade before, he declared them 'quite as sweet as partridge'. In 1887, a colonist described Aborigines on the Upper Darling catching rats by partially blocking the openings of their stick-nests, positioning themselves at each opening with a waddy or stick, then torching the nests to force out the rats.

Colonists and their stock destroyed many others. When Charles Sturt encountered these nests near the junction of the Murray and Darling Rivers in 1844, he too set fire to some and waited 'for the rats to bolt', in the hope of catching some. When Kenric Harold Bennett was searching for new sheep country in the Barrier Ranges in 1874, he torched more of these nests, curious to see if they were inhabited, unlike those he had encountered back of the Lachlan, and found up to a dozen rats. A journalist in Hay heard in 1887 that pastoral workers on the Bulloo and the Paroo hunted these rats. He reported that 'when the station hand gets short of meat he takes his gun and goes forth and sets fire to a mound, and when the rat jumps out he bangs them', which would have required a good shot. Still, many stick-nests survived in Carrypunday Swamp with each one, according to Davy, 'the resort of a large number of rats'.

As Davy described them, the stick-nest rats were usually sedentary—something modern scientists have assumed. Because of the complexity of their communal nests and the amount of work required to construct them, the rats would have occupied these nests for generations. Davy reported that the rats stayed in the swamp when it was dry or 'only ordinarily or temporarily flooded'. But when 'extraordinary high floods' on the Bulloo submerged millions of acres, turning the swamp into a 'vast estuary', some but not all the rats, which Davy described as 'surplus stock', cleared 'off before it'.

The Milparinka mining warden's telegram suggested that one instance of rats escaping floodwaters on the Bulloo was in January 1886. Rats may also have dispersed in mid-1886 when the exceptional winter rains saw the Bulloo flood, then briefly subside, then flood again, or in December 1886 when the Bulloo again flooded, or in February 1887 when the river rose 'higher than for years past'. It was this flood that Davy fixed on, perhaps oblivious to the earlier ones or judging them insufficiently big. He contended that the 1887 flood was the first since 1870 when rats 'were known to leave before the rush of waters'. If so, the rats had travelled the 500 miles to Hay and Balranald in three months, again averaging five miles a night, as they reportedly did at the start of 1886 on their way to Milparinka.

A series of later accounts from the 1920s are revealing. One is an interview with Sidney Kidman, Australia's most renowned pastoralist, and Harry Wheeler, a pastoral manager, then station owner. Asked whether he remembered 'the rat plague of 1885', Kidman responded at length, though about 1886. He recalled rats coming in millions from the Cooper, jumping down a tank and nearly filling it on Naryilco station in Queensland, and erasing the road from Cobham Lake to Mount Poole in New South Wales. Wheeler added that 'the rodents were all over Bulloo Downs and built nests about 18 inches high out of sticks'. Thomas Burston, who lived around Thargomindah in the early 1880s, wrote more. He recalled a 'migration at the Bulloo' when, 'having no cracks in the ground in which to hide', the rats 'built large mounds of sticks cut in lengths of about five inches under which to shelter'—a rare instance of a settler witnessing the construction of stick-nests. According to Burston, the rats were heading 'exactly opposite to the rabbit'—in other words, going south.

When Davy's letter appeared in 1887, Kenric Harold Bennett was already pursuing the identity of the 'far famed Bulloo Rats', which he thought of as travelling down from Queensland for six months, since the winter floods in 1886. Bennett was not sure whether these rats were the species he encountered back of the Lachlan in 1864 and wrote about in the *Queenslander*. He thought the earlier rats were smaller and darker than those of 1887 which he described as greyish and about a foot long including the tail, bearing 'a strong resemblance' to the grey-brown Norway rat, with its grey or white belly. Whereas Humphry Davy implicitly identified the rats in 1887 as a species of *Hapalotis*—the genus that then included the two types of stick-nest rat—Bennett did so explicitly.

Bennett wanted their identity determined by Edward Pierson Ramsay whom Bennett had visited again in Sydney, prompting them to write to each other as 'My dear Ken' and My dear Ned'. In March 1887, Bennett offered to send specimens, assuming Ramsay would—and should—want to examine them. 'If the Museum requires any examples let me know,' Bennett wrote, 'and I will try to obtain some.' A fortnight later, when the 'main body of the rats' had 'arrived and passed on in their southward march', and Ramsay had not responded, Bennett repeated: 'Do you want any "Bulloo Rats"?' When Ramsay finally assented, Bennett sent at least two. A month later, Bennett's curiosity was palpable. 'Let me know if you have received the rats,' he wrote, 'and if they were or are of interest.'

Bennett had been sending Ramsay specimens since they began corresponding in 1879, assisted by spirits, arsenical soap and collecting bottles provided by Ramsay. At first, caterpillars ate the legs, feet and beak of birds he tried to preserve. Then Bennett learned from a pamphlet about collecting written by Ramsay that he needed to brush these body parts with mercuric chloride and camphor. Bennett also gradually became more successful with bottling specimens. But while he took 'a good deal of trouble' with the rats—most likely, preserving them as soon as possible in strong spirits dashed with a little arsenic, having opened their abdomens so the spirits could enter their corpses freely—they arrived in a 'bad state', making them harder to identify.

By then, the arrival of the Norway rat in Australia was complicating classification by local scientists eager to identify new native species. In a series of articles on natural history for the *Sydney Mail* in 1873, Ramsay's predecessor, Gerard Krefft, acknowledged that he had sometimes paid 'a shilling for a rat with a very glossy fur and a long tail', thinking it might be a new native species, only to discover it 'was the progeny of a Norwegian rodent'. Other naturalists were similarly confused. In 1882 two members of the Royal Society of Tasmania identified a rat common in the island's north as an unnamed native species, dubbing it *Mus griseocoerulus*. A year later, they received an identical rat from a colonist in New Zealand's South Island who believed it was native there. Within a few years, it was clear all were the Norway.

Rattus rattus, which probably came to Australia with the First Fleet, created more confusion among scientists. When John Gould received one in 1863 that had been nesting in bamboo on the Sydney harbour front, he mistakenly decided it was a new native species and dubbed it *Hapalotis arboricola*. Edward Pierson Ramsay also blundered in 1882 after receiving a rat with long black hairs down its back and side found in a house near Wagga Wagga. While Ramsay recognised it was similar to Gould's *Hapalotis arboricola*, he identified it as another new native species, naming it *Mus tompsonii* after its collector. Just as *Hapalotis arboricola* was eventually expunged from the scientific record, so was *Mus tompsonii* when the Australian Museum received more specimens from Wagga Wagga that fitted Ramsay's description and were all *Rattus rattus*.

When Ramsay identified Bennett's specimens as *Mus tompsonii*, Bennett was unconvinced because he thought them a form of *Hapalotis*. Was *Mus tompsonii* truly 'the name of the beast,' he asked. 'Are you sure of the identity?' But Bennett was in no position to argue because he was the amateur, not the professional, and he was in Ivanhoe, not Sydney. When Bennett sent an account of the rat to Ramsay for presentation to the Linnean Society in Sydney—as long as Ramsay judged it 'worthy of being heard'—Bennett did not state that Ramsay had identified it as *Mus tompsonii*. Nor did he include a description of the rat, which might have raised questions about its identity. Instead he presented it without explanation as *Mus tompsonii* which meant that his paper, for many years, was largely ignored or read as an account of *Rattus rattus*.

Journalists filled this void, providing many descriptions of the rats that abounded in New South Wales in 1887. A correspondent of the *Riverine Grazier* wrote from the back of Hay: 'They somewhat resemble the brown rat, but are not identical, there are long hairs through the fur not found in the civilised rodent, the snout is a shade shorter, and the shape of the body slightly different.' The *Grazier* itself reckoned the rats had 'not the face of the European animal' but were 'doglike in appearance'. The Hay correspondent of the *Australian Town and Country Journal*, who saw several 'very large' specimens, described their 'visage' as 'somewhat different' to those he had seen in coastal towns. The newspaper's Burke correspondent identified them as 'pure water rats'.

Accounts of the rats found elsewhere also varied, if not always quite as much. For example, those in northern Victoria in the first half of 1888 were described as relatively light in colour and relatively dark, as brown and black. The most precise report distinguished them 'from their cousins that infest the sewers and cellars of our cities', with longer hair, more pointed noses, lighter colour and 'neither the cunning, nor the ferocity of town rats', so they were 'trapped with the greatest ease'.

Their size was a particular issue. In 1886, when he was working near Silverton, Rabbit Inspector Holding provided a measurement of one he thought particularly large: '16 inches from the tip of the tail to the nose'. Other observers relied on comparisons with introduced rats, usually *Rattus norvegicus* but sometimes *Rattus*

rattus, reaching very different conclusions. 'Somewhat smaller than the ordinary house-rat', reported the *Silver Age* in 1886. 'About the size of ordinary house-rats, but with thicker fur, lighter in colour, and shorter tails', reckoned the agricultural reporter of the *Sydney Mail* in 1887. 'Of enormous size', 'larger than the common or Norway rat' and 'larger than the ordinary house rat', ran reports from northern Victoria in 1888.

Most likely, more than one species abounded, which would help to explain the variant accounts of the appearance and habits of the rats as they spread across much of the continent. Consistent with Humphry Davy's account, some may have been stick-nest rats that originated partly from the Bulloo and, prompted by the floods of 1886 or 1887, travelled far and wide. Thomas Burston's reminiscences of rats building stick-nests as they travelled support this identification. The dearth of other such accounts does not negate it, as greater stick-nest rats on the Franklin Islands have generally done without stick-nests—opportunistically using burrows of other animals or hiding beneath boulders or dense vegetation.

The prime species, including that in northern Victoria, was almost certainly the long-haired rat—and not just because a wealth of evidence identifies it as Australia's pre-eminent irruptive rodent. This identification is supported by the report of the rats' 'long hairs through the fur'. Skeletal evidence provides corroboration. The mummified remains of a long-haired rat have been found in the wall cavity of Garnpung Homestead which was built in the mid-1880s south of what is now Mutawintji National Park, an area in far western New South Wales where rats were only recorded in 1886 and 1887. Some of these long-haired rats may have originated from the Carrypunday Swamp, where they are still to be found.

CHAPTER 13

If Rats Kill Young Cats

William Ranken's *Dominion of Australia* excited immediate attention in 1874 because of its focus on Australia's climate as 'a most capricious tyrant, destroying at uncertain intervals what it has reared in a few milder seasons'. Ranken wrote from personal experience, having spent several years managing a station in Queensland. But Ranken also maintained that life could be 'not only agreeable but profitable' if colonists were careful in selecting land on the fringes of Australia's 'central oven'. His choice, a few years later, was Tongo Station on the Paroo River in the far north-west of New South Wales. Its 176,000 acres with almost 35,000 sheep almost ruined him, before an inheritance allowed him to buy another property further south.

The rats had just reached Tongo at the start of 1887, when a boundary rider went to inspect the station's fences and repair any broken sections, leaving six kittens in his hut on the Paroo. While generally just a series of waterholes, with no flow between, the Paroo was rising fast that wet summer and would soon stretch fifteen miles across the plains. When the boundary rider returned, the kittens were dead and, the *Central Australian* newspaper reported, he blamed the rats.

Lake Eyre in northern South Australia was considered to be the centre of Australia from the start of the twentieth century; Uluru, then known as Ayers Rock, in the Northern Territory, replaced it in the mid-twentieth century. The *Central Australian* was published much earlier, further east—from 1867 in Bourke on the Darling River in New South Wales. Only a few issues exist from its twenty-five or more years in print. The issue including the report about the killings on Tongo Station does not. But its account survives because newspapers across the continent reprinted it. They did so because the rats, if carnivorous, might stop rabbits devastating the land. As the *Central Australian* put it, 'If rats kill young cats, they will also kill rabbits.'

The crisis in New South Wales had been growing as the rabbits spread despite immense public and private expenditure. As military analogies became commonplace, one horrified settler observed: 'The war against the rabbit was entered upon with a light heart, which rose from ignorance as to the nature of the undertaking... The country never for a moment imagined that it was creating a standing army larger and more expensive than our permanent military force.' While Kenric Harold Bennett considered that anyone with common sense would know that controlling, let alone eradicating, the rabbits, was 'impossible', most colonists had no idea that killing millions of rabbits would have scant impact. In late 1885, the rabbits were estimated to be travelling north through

New South Wales at a rate of one hundred miles a year, putting them within reach of Queensland. In January 1886, there were reports—immediately disputed—that some had been shot across the border. Within a year the rabbits had reached a hundred miles into Queensland.

The future looked even bleaker because of the rabbits' fecundity. A recurrent question was what would happen if a pair of rabbits bred unchecked in favourable conditions, free of predators, and their offspring were similarly fortunate. A famous answer by the English zoologist, Thomas Pennant—predicated on the rabbits breeding seven times a year and producing litters of eight—was 1,274,840 progeny in four years. As colonial mathematicians assumed the rabbits' new environment engendered bigger litters more often, their estimate ranged from two million in two-and-a-half years to fourteen million in three years to 1500 million in four years.

While most landholders employed trappers to kill the rabbits, many tried poisons such as strychnine and arsenic. Some erected stone walls and net fences. Others experimented with ferrets, dogs and cats or looked for a marsupial predator, hoping they would be much cheaper and more effective than employing trappers. George Hebden of Gogeldrie Station in the Riverina proposed a scheme involving a million cats. Half were to be *Felis catus*, the introduced animal that could be bought for a shilling or two in the colony's towns and cities. The other half were to be *Dasyurus viverrinus*, the tiger cat or eastern native cat, now usually known as the eastern quoll, a marsupial that would become extinct on the Australian mainland in the 1960s but survive in Tasmania and, after fifty years, be reintroduced to the mainland in small numbers.

Hebden expected these cats, both introduced and native, to

bring the rabbits under control since, even if the rabbits constituted just a quarter of the cats' food, 'that would amount to a quarter of a million rabbits per day' or 'over seven million a month'. Hebden recognised the cats would also devastate native birds that protected the colonists' crops by killing insects, but maintained the settlers 'had to risk something'. Having 'destroyed the equilibrium of nature' by releasing the rabbits, colonists needed 'to restore that equilibrium, and even go a little further by introducing small carnivores in large numbers'.

The population of eastern quolls in the Riverina had plummeted, at least partly due to the colonists' profligate use of strychnine to kill the dingo. When Hebden sought to put his scheme into practice, he consequently looked for quolls from elsewhere. He was assisted by another of the quolls' advocates, Frederick Campbell of Yarralumla, whose house has become part of that of Australia's Governor-General in Canberra. In mid-1885, Campbell took out advertisements in local newspapers seeking trappers. They caught 780 quolls around Goulburn, Queanbeyan and Bungendore, then sent them by train to the Riverina, the first big attempt at relocation of a native species in Australia.

Some of the quolls died through overcrowding, others from being boxed up wet. But 700 survived and at an average cost of two shillings and a total cost of £100—less than the cost of employing a rabbiter for a year—Hebden declared them a bargain and made much of their efficacy. Then, he turned to the introduced cat, releasing 200 to 300 in the winter of 1886, which he admitted was a failure. The problem, he maintained, was not the cats' capacity as rabbit-killers, but that surrounding stations were more heavily infested than Gogeldrie, so his cats were lured away by even more abundant food and died in rabbiters' traps. He was sure that, had

Gogeldrie been an island, 'there would not now be a rabbit left upon it'. When he acquired more cats, he declared them a success, though, general experience would suggest, they would have killed many of Gogeldrie's rabbits but failed to control them.

The government also contributed when it had at least 200, perhaps 400, quolls trapped just north of Sydney, then sent them in March 1886 to Bourke by train in small cages. The *Queenslander* was appalled that anyone would extend the range of such ferocious animals and was delighted when at most 100, perhaps just thirty, survived the journey. 'We do not believe in introducing one plague for the purpose of mastering another,' the *Queenslander* declared. 'Once let the ferocious little *dasyurus* get the upper hand and they will not go hunting for rabbits...when they can more easily obtain the eggs and young of birds.' The Melbourne *Herald* concurred, albeit on the basis that the animals sent inland were introduced cats, not native ones. In an editorial ironically titled 'Pussy to the Rescue', the *Herald* wondered whether, just as several other introduced species had become more destructive in Australia, so would *Felis catus*. 'They might develop into small tigers,' it warned.

The destination of this consignment was Tongo Station where William Ranken must have been happy to take part in this experiment in rabbit control. Rather than being carried by cart from Bourke, some reports improbably suggested the quolls were herded like sheep, prompting one newspaper to dub them 'Queer Travelling Stock'. While the kittens in the boundary rider's hut may have been domestic cats, which the boundary rider was raising either as pets or rabbit-killers, they might also have been progeny of the quolls sent to Tongo.

William Reid of Tolarno Station—a vast property near Menindee with a frontage of fifty miles on the Darling and extending back eighty miles from it—tried only *Felis catus*. Because of the rains, he could buy cats in Adelaide, then ship them by steamer along the Murray and the Darling, still the best means of transport into the inland when these rivers were navigable. Reid began in October 1886 with 300 cats, which, by one account, immediately 'rejected rabbit-hunting as unprofitable' and, rather than straying onto other stations, invaded Menindee. But Reid's acquisition of another 100 that December, and 200 more in February 1887, suggests he considered them a success or at least an experiment worth trialling on a bigger scale.

Frank Powell, the managing partner of Thargomindah Station on the Bulloo, also contributed to this quantum leap in the number of cats in the bush. After rabbits were first sighted on the Bulloo late in 1887, gaining easy access from New South Wales with the Queensland Government's rabbit-proof fence not yet complete, Powell advertised in February 1888 for 'any number of cats' delivered to his station. He offered to pay two shillings and sixpence per head—the high end of the price paid by pastoralists wanting cats as rabbit-killers. The result was the first reported plague of cats in the Australian outback. A much-reprinted account by the Thargomindah correspondent of the Brisbane *Courier* in July 1888 was of 'thousands upon thousands of cats' in 'very poor condition...advancing from the north' and 'infesting the bush'. It was 'supposed'—ignoring Powell's role—the cats came in search of the rats 'due to the scarcity of food'. According to a slightly later report, 'immense swarms of rats swept over the land, and in their track followed crowds of domestic cats...which preyed on the wandering rodents'.

Cat plagues occurred elsewhere. A worker on Barenya Station near Hughenden, about 500 miles due north of Thargomindah, recalled how 'an army of cats followed the rats'. He counted 'as many as thirty cats one evening at a lamb marking camp'. On other occasions he 'noticed scores of cats' and heard similar reports from neighbouring stations. A young black cat still carried a mark of its recent domesticity, 'a ribbon around its neck'. These cats soon succumbed to a disease generally believed to have resulted from the cats 'mixing with the rats'. They died 'with their heads a mass of blotches' and, long afterwards, 'the creeks and billabongs were littered with the remains of dead cats'.

The focus in the wake of the boundary rider's story from Tongo Station was very different. It was on rats as kitten killers and whether they might kill rabbits too. If the rats would, they offered great advantages as a control because they were already where the colonists needed them, did not have to be purchased, were also in plague proportions and, though colonists did not fully appreciate it, could out-breed the rabbits. The journalist who first reported the killings on Tongo predicted that, just as colonists sometimes prayed for drought-breaking rains and thanked God if they came, so they might thank Providence for the rats.

This welcoming of the rats—this delight in a plague—was remarkable given the colonists' fear of them, shaped by the European experience of the English and the Norway rats. No other animals were 'a more formidable enemy to mankind' or 'more injurious to the interests of society', the great English wood-engraver Thomas Bewick declared in his celebrated *History of Quadrupeds* first published in 1790. Yet the story from Tongo also had great European resonances—above all, with the most popular English version of *The Pied Piper of Hamelin*, written by the poet Robert Browning in 1842. While the tyranny of cats over rats was the stuff of hundreds of stories, from *Aesop's Fables* in classical Greece to *Piers the Plowman* in medieval England and beyond, Browning's second verse inverted this order of things:

> Rats!
> They fought the dogs and killed the cats,
> And bit the babies in the cradles,
> And ate the cheeses out of the vats,
> And licked the soup from the cooks' own ladles…

English accounts of rats killing rabbits also abounded, along with stories of rats killing lambs, pigeons and poultry. Some of these accounts emphasised the remarkable capacity of rats to reach their prey, others their rapacity. A typical story involved more than a hundred rabbits living in burrows beneath a house in London where they were being bred. When Norway rats in the sewers smelled the rabbits, they broke through from the drains and killed all but five of the rabbits, consuming the bodies of some, 'leaving the heads and legs…hanging merely by bits of skin', and eating the heads of the others, leaving 'only a portion of their skins, feet, and ears'.

Such stories were intended to shock and entertain. Hyperbole was assumed, sensation expected. The accuracy of these stories was rarely an issue since little or nothing turned on their truth. The story from Tongo Station shared many of these characteristics. In the language of the day, it was not just a 'yarn' or entertaining story, but a 'tough yarn' which, the colonial writer Dan Deniehy once observed, might be told on April Fool's Day, was ripe for a new Baron Munchausen, and could have been recounted by Scheherazade on the Thousand and Second Night.

Some form of verification was needed. It came from a local

landholder, Richard Hodnett, who informed the *Central Australian* that he had investigated the boundary rider's story and vouched it was true. Hodnett's identity as a member of the local pastoral protection association and the executive of the colony's agricultural society, who often spoke about rural issues, ensured he was taken seriously. His message of hope was that the rats had 'invaded the rabbit country, taking possession of their burrows and scattering them about, giving infected districts a chance'.

Other colonists had considered this idea since the rats began spreading. 'May I suggest,' wrote a reader of the *South Australian Advertiser* in February 1886, 'some of the Government rabbit parties in the north should proceed further north and catch alive a few hundred of the rats that we read about, and put them on the rabbit infested country? I feel certain the rats would kill all the young rabbits, if not the old ones.' South Australia's Pastoral Board concurred after finding the colony's far north-east 'swarming with rats' but 'few rabbits' that April. One explanation was that 'scalping parties had done their work well'; another was that 'rats disposed of many of the rabbits'. In western Queensland, where the rats were reckoned to be 'almost as destructive as rabbits', there was a 'strong belief' the rats could do colonists 'some service in destroying the invader'.

Rabbit Inspector Holding reported the same from north-western New South Wales. At Cobham Lake, the rats appeared 'to be clearing the rabbits out—if not destroying them, at least hunting them out, as they invaded all the burrows and holes along their tracks'. Outside Broken Hill, the rabbits had been out of control until their burrows were 'invaded by rats'. Thereafter, Holding saw 'no signs of rabbits except the old dunghills and scratches'. When he checked with local Wilyakali people, they told him of 'one or more instances in which they had found the carcasses of large-sized rabbits partly devoured, and the tracks of rats thick around them'. European trappers similarly 'had often found rabbits in the traps partly devoured by rats; rat tracks only being visible about them'. Holding concluded it was 'almost a certainty' the rats would find 'such timid and defenceless creatures' as the rabbits 'easy prey', especially in their burrows. All who knew the rats were 'aware of their rapacious nature'.

Other inspectors disagreed. One near Broken Hill reported that, while rats and rabbits were to be found in the same warrens, there were no signs of the rabbits being destroyed. Because the many rats caught in traps were 'usually eaten by the others', he thought the rats 'would probably eat young rabbits if they could find them', but concluded that the rats failed to do so. Another inspector near Silverton found the rats did 'not seem to care to live in rabbit burrows' but preferred to make their own. When he consulted Aborigines, they reported the rats did not interfere with either adult rabbits or their young. When he asked local trappers, both black and white, they agreed 'the rabbits did not mind the rats in the least'. When he dissected several rats, he found no traces of any animals in their stomachs.

Humphry Davy concurred when he wrote to the *Riverine Grazier* in April 1887 about the rats from the Carrypunday Swamp: 'These rats are non-carnivorous and live on roots and herbs. They are most harmless little creatures, and all through this country they are the prey of everything, and already I have seen hundreds of them dead.' Davy thought it 'as likely that rabbits would kill them as that they would kill rabbits, and there is not the slightest chance of rabbits becoming less on their account'. Having 'left their natural haunts', he expected the rats around Balranald and

Hay would 'soon become extinct in these districts'.

Journalists divided. One reported: 'Something must be at work for I hear no complaints about the rabbits, and they are now undoubtedly on the decrease.' He credited the rats with this 'encouraging work'. Another was hopeful the rats would 'keep the rabbits down'. A third wanted to hear from 'station people, practical men', expecting they would deliver a different verdict. A fourth predicted that, even if the rats were eating rabbits, it would not stop the rabbits 'marching to victory'. A fifth identified the rats as 'vegetarian'—a term only recently in general usage following the formation of the Vegetarian Society in England. In case its meaning was not understood, this journalist added: 'They do not eat meat.' A sixth engaged in a typically disparaging Islamic analogy: 'It would seem to be a pretty bold policy on the part of a rat—even on the part of a Queensland Aboriginal rat to tackle a full-size buck rabbit, but then it must be remembered that these rats are like Mahometans—there is no fear of death before their eyes, though we are unable to explain the reason for it.'

This interest and confusion were all the greater because the encounter between the rats and rabbits was new—a meeting even more remarkable because of how the rats and rabbits came from different directions, surmounting great natural obstacles. While the rats headed south into New South Wales and South Australia, the rabbits headed north. Little realising that all Australian rodents have at least some ability to swim, the roving agricultural reporter of the *Sydney Mail* recorded with amazement how, 'by diving and swimming', the rats 'were enabled to cross all the floodwaters'; 'numbers were drowned through the wind blowing from the direction in which they were swimming, and so causing them to breast a strong current', leaving many dead on the shores of the

Menindee Lakes; but 'no matter what the difficulty they never turned back'. Other colonists expressed similar amazement at how, as they saw it, the floods had also taught rabbits to swim, they took to the water with ease, bounded across the plains and climbed trees with aplomb, so they sometimes seemed like a new species, very different to their English ancestors.

Humphry Davy was behind several accounts of the rats' rapid evolution. After talking to him in September 1887, Hay's *Riverine Grazier* identified the rats as 'quite a countenance to the Darwinian theory'. It elaborated: 'In the Bulloo district they build mounds of grass and live in these mounds in communities. But since their advent into Riverina, when they were driven out of the Bulloo regions by the floods, they have developed new traits: they now dig burrows. In their old habitat, they were no harm; but here they are developing mischievous traits.' The Hay correspondent of the *Australian Town and Country Journal* echoed: 'The Bulloo rat has, like every immigrant animal which comes here, changed his habits. He was satisfied with a mound on the Bulloo. But here he goes in for burrows, perhaps incited and encouraged in doing so by the villain rabbit.'

The rabbits' adaptability excited much more attention since, if they could learn to climb, the colonists' rabbit-proof fences were set to fail. The Menindee postmaster—another public official confident of his capacity to observe and theorise about his new environment—fueled these fears when he filed a special report in November 1886 with the Rabbit Branch of the Mines Department. It was prompted by the postmaster's discovery of fourteen full-grown rabbits in two trees where, he figured, they had spent 'ten days to a fortnight, and were quite at home—some of them perched on branches twelve feet up, or camped on the fork of limbs'. He believed these rabbits 'had been driven to high ground, which eventually was covered by the flood, and being thus surrounded took to the trees'. Australia's greatest writer of the era, Henry Lawson, was typically concise in his first book, published a few years later. The bush, Lawson observed, was turning the rabbit 'into something like a cross between a kangaroo and a possum'.

The idea that the rat might be the colonists' saviour was reiterated by H.E. Kingsmill, an agent for pumps and hot-air engines in the Northern Territory who visited Sydney in late 1887. Kingsmill was unaware that the rat had already been identified as a possible means of rabbit control. He did not realise that it abounded in New South Wales that year. Instead, he proposed its introduction from north Queensland, 'its natural habitat', where he had encountered 'large numbers in both Normantown and Burketown'. Kingsmill was 'convinced from watching its nocturnal movements' that the rat was 'capable of taking a very firm grip on its prey and inflicting at once a fatal bite. Then numbers gather round and share in the feast, which is quickly disposed of.' While he did not suggest it could kill adult rabbits, he was sure it 'would be death to all the young rabbits'.

Even if stick-nest rats were part of the great irruption, they could not control the rabbits because they ate only plants and insects. The long-haired rats were different. While primarily herbivorous, many reports and studies, starting with Thomas Wall and John Gould, have identified them as sometimes carnivorous, readily eating carrion and possibly killing other small animals. As a result, the long-haired rat probably ate many rabbits killed by the trappers. It may also have killed and eaten young rabbits in their burrows and killed the boundary rider's kittens on Tongo Station. But the rat could not slow or stop the rabbits' spread, as the Bulloo and its swamp illustrate.

Here, more than anywhere in colonial Australia, was an area renowned for its rats, which also made it ideal for rabbits, offering abundant feed in the worst of seasons and perfect shelter in the surrounding sandhills. Although the local Marsupial Destruction Board extended its operations to rabbits in 1888 and soon employed thirteen parties to destroy them, Bulloo Downs became the worst-infested property in Queensland. The colony's surveyor-general, Archibald McDowall would often see as many as fifty rabbits when he visited. At other times, there were too many to count and their droppings were so thick McDowall was reminded of a yard where sheep had been penned overnight. When the station's cattle were in terrible condition due to lack of grass, the rabbits were fat as seals, thriving on bark and twigs that they stripped from bushes and trees. While the rats survived, with the Bulloo Overflow providing their most important refuge in New South Wales, Bulloo Downs was home to one-quarter of Queensland's rabbits in the twentieth century.

CHAPTER 14

Hatred and Suspicion

Many species fueled the colonists' realisation that the country was beyond their control. Locusts—eventually identified as three main native species—were their most feared 'insect enemy' because they spread in prodigious numbers, devastating crops and pasture. The exotic daisy known as Bathurst burr, which arrived in the wool of sheep and the tails and manes of horses shipped from South America, was the settlers' 'enemy in wool'. Of no use as feed, it smothered other plants, while its burrs damaged fleeces and were very difficult to remove. Many pastoralists feared its spread could put an end to wool growing.

Reports of miscegenation in the outback creating new hybrid species reveal that colonists not only enjoyed 'tough yarns' but also feared they had landed in a place where anything could happen. These reports all involved the rabbit, as settlers were amazed and appalled by its 'phenomenal fecundity'. Some involved the bilby, the Australian species most often likened to it. Many colonists reputedly believed that the bilby itself was the product of the rabbit mating either with a bettong, or rat-kangaroo, or with the common bandicoot. 'The bilby is not a cross between the kangaroo-rat and rabbit,' a Sydney columnist felt compelled to explain: 'Absurd to speak of rabbits and bandicoots "crossing". Placental and non-placental mammals *can't* cross.'

As other stories proliferated, one report from back of the Lachlan involved the rabbit breeding with a bilby. Another, of a rabbit mating with a cat near Horsham in northern Victoria, led two government officials to investigate. One was the local crown lands bailiff, the other a lands officer. While their positions required no zoological expertise, their reports were said to put 'the correctness of the statements made beyond all doubt'. According to the bailiff, one of the hybrids was a bright orange-yellow, the other pure black. Their 'chief resemblance to the rabbit' was in their 'short and turned up' tails. Although they could walk like cats, they were 'evidently awkward' when they did so and, 'on any attempt being made to accelerate their speed', adopted 'the peculiar hop of the rabbit'.

Melbourne's foremost scientists became involved when a local publican bought two more of these 'cat rabbits', then donated them to the city's zoo where they were exhibited as 'supposed hybrids'. When challenged by a visitor, the zoo's director Albert le Souef responded that the animals 'partook of the characteristics of the cat and the rabbit in a remarkable degree'. The director of the National Museum of Victoria, Frederick McCoy, concurred because of the creatures' 'short upturned rabbit-like tail, disproportionately large hind quarters and legs, small front legs, and hopping movements'.

The long-haired rat, photographed by Angus Emmott, outside Longreach in central western Queensland during the plague of 2010–2011.

The earliest image of the long-haired rat, published in London in 1854 as part of John Gould's *Mammals of Australia*. It depicts the one specimen that Thomas Wall, the natural history collector on Edmund Kennedy's expedition to the Barcoo River and Cooper Creek in 1847, succeeded in bringing back to Sydney.

One of Ludwig Becker's watercolours from the Burke and Wills expedition, painted on 9 March 1861 at the site in the far north-west of New South Wales his companions called 'Rat Point'. The two dingoes in the foreground would have been attracted by the long-haired rat.

Ludwig Becker's watercolour of the long-haired rat, painted on 23 February 1861, at 'Rat Point'.

Kenric Harold Bennett in the heart of scientific Sydney—the boardroom of the Australian Museum. Ornithologist and oologist Alfred North is at left. Bennett is second from left, followed by an unidentified man and then the museum's curator, Edward Pierson Ramsay.

The first depiction of the rat en masse—one of Arthur Pearse's illustrations for 'The Adventures of Louis de Rougemont', published in London's *Wide World Magazine* in February 1899. De Rougemont and his 'ever-faithful' dog Bruno find safety up a tree while 'even the biggest kangaroos' fall prey to the long-haired rat.

The long-haired rat as letter-winged kite food—a photograph taken by the naturalist Sidney Jackson while on Davenport Downs on the Diamantina River in Queensland in 1918. After killing the rats, Jackson fed them to the kite chicks.

Wangkangurru man Jimmy Naylon Arpilindika caught the long-haired rat for zoologist H. H. Finlayson around the Diamantina River in South Australia's north-east in December 1931. Finlayson, who looked on Naylon as the 'rat boss' and lauded his 'amazing skills as a hunter', took this photograph.

When H.H. Finlayson first studied the long-haired rat in 1931, he was primarily interested in the oolacunta, *Caloprymnus campestris*. Finlayson became famous for rediscovering this desert rat-kangaroo, but it soon became extinct. This remarkable photograph, 'The Oolacunta As He Appears At Speed', was taken by the one-handed Finlayson while running it down on horseback.

During the plague that peaked in 2011, Angus and Kate Emmott of Noonbah, outside Longreach in Queensland, were exceptional in trapping the long-haired rats in their house and releasing them—well away from the house. This photograph, by Angus Emmott, shows one night's catch from their kitchen.

When one of the animals died, Baldwin Spencer, the foundation professor of biology at the University of Melbourne, supervised its dissection. In a widely republished report, providing precise anatomical detail, he pronounced it a cat.

There were also several sightings of the creature that the Wergaia people of north-eastern Victoria called the bunyip. Part of the Dreaming of several Aboriginal groups in south-eastern Australia, it excited particular attention from colonists sometimes set on ridicule, sometimes eager to locate and identify it. While William Ranken dismissed the bunyip as 'some animal, probably an inland seal, now extinct', which Aborigines 'in their ignorance learned to dread', it entered settler folklore as 'the one respectable flesh-curdling horror of which Australia can boast'. By most accounts, the bunyip was an amphibious creature that lived in lagoons and deep waterholes and emerged only rarely at night.

Officialdom again loomed large when a bunyip was reported on the Lachlan River outside the township of Oxley. The local police sergeant led a posse of twenty armed men to catch or kill this creature. They found a common duck. Other reports, such as the sighting of a seal-like animal, about three-feet long, on an anabranch of the Lachlan, were not confirmed or disproved. According to the gentleman of 'unquestionable veracity' who spotted this animal in 1887, it dived into the creek once aware it was being watched, then emerged several times. The journalist who reported this episode thought it 'only fitting' one of these 'semi-mythological animals' should appear when the big rains gave rise to 'all sorts of stories'.

As conflicts over the land took many forms, stories about rabbiters made outback Australia appear all the more out of control. A common assumption was that the trappers were so greedy that rather than trying to exterminate the rabbits as quickly as possible they sought to perpetuate them. One allegation was that the rabbiters always left some behind, especially pregnant does, to sustain the rabbiters' 'good line of business'. A variant was that the trappers tried harder to destroy the rabbits' predators than to kill rabbits themselves. Another was that the trappers viewed the spread of the rabbits as a 'God-send' and the 'making of the country, with plenty of work and good times for working men'. Some trappers were said to be so 'warped in this matter' that they deliberately introduced rabbits into uninfested areas.

Rabbit inspectors, whose annual salary of £315 was a vast increase on what many had earned, excited similar accusations, outrage and derision. Most were said to owe their posts to the patronage of powerful relatives in Sydney, with Kenric Harold Bennett probably depending on his uncle William Brodribb who, after being elected as a conservative member of the colony's Legislative Assembly, had been appointed to its Legislative Council. Far from their work requiring any skill, the inspectors were identified as 'scalp-receivers', who simply oversaw the counting of those killed by the rabbiters. They were accused of bringing rabbits into their districts and freeing them to extend their 'lucrative billets'.

Pastoralists were accused of other abuses. It was said that, to recoup the rabbit levy imposed by the government, some charged extortionate sums for allowing rabbiters to work on their properties. Others were accused of deliberately employing incompetent trappers or even trying to extend rabbiting because of the income it brought them. As one station manager said of such a squatter, 'The more men he had on the run, the larger profit he made on the sale of stores, and therefore when a man caught too many rabbits a quarrel was picked with him, and he was dismissed.' When some squatters were prosecuted for failing to control rabbits on their runs, the key to the conviction of one near Ivanhoe was that he imposed such excessive costs on rabbiters that just ten would work for him when his station required thirty.

The rabbiters attracted most criticism, fueled by class prejudice, when their income rose sharply as the rabbits multiplied. While speculation and exaggeration were characteristically rife and hard figures almost impossible to come by, there was outrage at the rabbiters earning three to four times as much as ordinary station hands, eclipsing shearers who were much more skilled workers, and sometimes earning more than station overseers. The greatest concern was that these men refused to treat their masters with respect, creating 'insubordination on the stations'.

An array of stories—perhaps all apocryphal—had Aboriginal rabbiters using their unprecedented earnings to buy buggies and horses. In one story from Hay, the rabbiter was duped into acquiring a vehicle and horse 'in the last stages of dissolution'. In another from Silverton, a rabbiter ordered 'a gorgeously-finished dog-cart' which came with an extravagant harness decorated with bells and was 'embellished with paintings of rabbits scampering off to their burrows, pursued by a chorus of piccaninnies, gins and dingoes'. The prospect of an 'Australian black' planting his bare foot on the steps of a new carriage and vaulting onto its cushioned seat was treated by commentators as a risible inversion of the social order. But they typically judged it better that Aboriginal men should be acquiring good horse flesh and flash vehicles than drinking their wages as rum.

Other stories, imbued with even greater social panic, fixed on white trappers. Their enthusiasm not just for driving two-horse buggies and traps but for driving horses four-in-hand scandalised pastoralists who considered such transport their preserve. The trappers' purchase of cigars and alcohol was taken as proof that they were being overpaid. According to one report, a rabbiter in Silverton bought a supply of cigars for one shilling each—a prodigious sum when the annual rent for half the land in western New South Wales was one penny an acre. For trappers in Wilcannia, champagne suppers were said to be a 'recognised institution', though not an 'every day occurrence'. One of these trappers reportedly put a cheque for over a thousand pounds on the bar to pay for a sixpenny drink.

The trappers' alleged propensity for violence stoked fears that other working men would follow suit. One commentator deplored the 'demoralising tendencies' of 'the usual scenes of personal violence and brutality' that accompanied the rabbiters' 'carousing debauchery' when they went on sprees in town. While Kenric Harold Bennett was based in Ivanhoe, he railed against the 'wantonness' of these 'brutes' who, perhaps simply looking for amusement but more likely wanting to spite him, frustrated his attempts at egg collecting by torching nests he had been watching. 'These fellows are the bête noir of the naturalist as nothing is sacred from them,' he fumed. 'Their creed is kill or destroy everything. There are a great many of the larrikin class.'

These tensions erupted outside Tilpa while Bennett was there. When the manager of Marra Station refused to pay a rabbiter, the rabbiter's mate threatened legal proceedings to recover the money. 'I'll kick you,' the manager responded. 'You can do that if you like,' the rabbiter declared. The manager rushed the rabbiter, hit him in the eye and felled him. When the man got up, the manager felled him again. When back on his feet, he brought out a pocket-knife, and, when rushed again, he stabbed and killed the manager. A jury in Sydney, in dismissing the rabbiter's prosecution for murder but finding him guilty of manslaughter, recommended mercy given his 'great provocation'. The judge sentenced him to ten months' imprisonment, but without hard labour.

The beginning of unionisation was a response to such class conflict, but also heightened it. Kilfera station outside Ivanhoe, where Bennett had observed the rats removing all traces of the sheep's hoof prints, was a prime site of confrontation. It was notorious among shearers for its owners' attempts to abuse provisions, allowing them to reduce the shearers' payments if sheep were badly shorn. When members of the new Australasian Shearers Union went on strike in 1887, some pastoralists laid charges against them and tried to recruit non-union labour. But Kilfera's manager, anxious to get its fine merino fleeces to market before the Bathurst burr spoiled them, negotiated with his 200 shearers. When they returned to work, having failed to secure their demands by some accounts but having triumphed by others, there was for once a show of 'amicable feeling'. The manager helped to stage an entertainment in Kilfera's prize-winning £4000 woolshed where a group of shearers and wool sorters, styling themselves as the 'O.K. Kilfera Minstrels', performed comic songs and ballads.

The rabbiters had particular reason to maintain the conventional European antipathy to rats and look on them with 'consternation and unlimited profanity'. If the rats ate rabbits in the men's traps, as many believed, the rabbiters lost out because they were paid for the number of rabbit scalps that they brought in. As many rats also beat the rabbits into the men's traps, they further reduced their incomes. To their horror, some trappers caught more rats than rabbits. A few allegedly responded by engaging in fraud, attempting to pass off rat skins as rabbit.

Kangaroo trappers, who were also paid per scalp, reputedly engaged in similar frauds. They were said to produce bags of 'dummy scalps made from kangaroo skin' with the stitching 'so neat as to be almost undiscernible except to an experienced person'. They were also said to 'remove the scalps of young kangaroos, with the ears attached and set them free again', leaving the kangaroos not just to breed but to do so exceptionally well, just as bald men were renowned for having unusually big families. Rabbiters were accused of making two animals out of one by skinning a rabbit and treating its fur and its carcass as if each came from different animals. They were said to make three animals out of two by even more complicated cutting and stitching. Another of their alleged schemes was to cut the unborn young out of pregnant does and seek payment for the entire litter as well as the adult.

The public response typically divided along class lines. On one side, there was outrage and expostulation; on the other, the rabbiters' actions were cast as trivial compared to those of colonial businessmen who sold underweight goods. The Sydney *Bulletin* observed that such abuses were regularly practised on a much larger scale by 'grog-watering, grog-making merchants' whose 'dodge' was to 'sell pint and quart bottles which are invariably under the nominal measures—three of the nominal liquid pints actually amounting to no more than two standard ones'. But while the rabbiter who made 'small profits' experienced the weight of the law, nothing was done about the 'other fellow, who is usually a man of much wealth and influence as a pew-holder'.

The courts became a forum for investigating these claims after an inspector discovered two men on Weinteriga Station near Menindee preparing to pass off rat skins as rabbit skins in January 1887. 'The rats about Menindee must be very large or the rabbits very small', observed the *Sydney Mail*, 'and the rats up there must be curiously like rabbits, or the rabbits must be curiously like rats, or else the fraud would scarcely have been attempted.' The similar colour of the rat and rabbit skins probably tempted them to do so. Justice Backhouse, who heard the case in Wilcannia, was as affronted by the rabbiters' actions as he was keen to deter other trappers from emulating them. He declared he would have liked them to have been caught after they had committed their fraud since he then could have gaoled them to five years. As it was, he sentenced them to two years' hard labour, the maximum penalty.

The second case involved three men on Gol Gol Station near Lake Mungo accused of trying to pass off four rat skins as rabbit skins. According to the prosecutor, the trappers had no excuse for engaging in fraud as they had been earning good wages when, prior to the rabbits' spread, they often had made no money at all. The prosecutor also made much of 'the curse the rabbits had brought on the country'. His prime witness was the local rabbit inspector who testified that the difference between the rats and rabbits was 'very considerable', so it would have been 'impossible' for the trappers to have overlooked it when they sought payment for the rat skins.

The accused responded that they had not skinned any rats and had presented them for payment accidentally. As part of their convoluted defence, they claimed the rats belonged to another trapper who had shared their camp but was not in partnership with them. It just so happened that, when preparing to move elsewhere, they had taken this trapper's rat skins from the wire where they were drying, put them in a bag containing rabbit skins, and forgotten they were there. The men had been asked by the trapper's wife to present the bag for payment since her husband was in hospital. Had they waited until he recovered, the skins might have been eaten by weevils.

The plausibility of this defence depended on the accused being able to explain why the fourth trapper should have been drying rat skins. This trapper, who was the brother-in-law of one of the accused, testified that he had caught the rats under contract to a local businessman who hoped to sell the skins. The businessman corroborated that he had asked the trapper to secure several dozen rats and had sent a parcel of skins to the city to test their value. This claim was improbable when most rabbit skins in New South Wales were unsaleable due to oversupply, and hence were destroyed. The sale of rat skins was also unlikely given the European assumption that rats were good for nothing, even though rat skins were used in France to make ladies' gloves, as a substitute for kid, and hats, as a substitute for beaver, and skins of the much larger Australian water rat would later become substitutes for pelts of the American musk-rat.

The prosecution of the men from Gol Gol station was the first criminal case heard in Balranald's new Italianate courthouse, which had opened with a champagne luncheon for local grandees. The Crown took the case so seriously that it brought in a Queen's Counsel from Sydney, while the trappers were represented by a solicitor from Hay. The novelty of the occasion, and interest in the case, drew 'several ladies from the country and town'. The presiding judge again was Justice Backhouse, as he went on circuit through far western New South Wales. He found that the inspector's testimony about the difference between the rabbits and the rats suggested the trappers had been engaged in fraud. He found that the men's continued employment on Gol Gol suggested their innocence. The overall impression was 'very even but rather in favour of the accused'. The jury found them not guilty.

Part III

Decline

CHAPTER 15

On the Shores of Lake Killalpaninna

The Lutheran mission at Lake Killalpaninna on Cooper Creek was like a small outback township, but with no hotels. Its dominant structure was a church with a twenty-metre-high bell tower and a fourteen-metre-long nave, built of mud bricks made by Aborigines at the mission. Other buildings included three houses for missionaries, a schoolhouse, two stores, a bakery, workshops, four cottages for Aboriginal couples, married by the Lutheran pastors, and a dining room and kitchen for the entire Aboriginal community. While its members at first were primarily Diyari, since the mission was on their land, it became home to members of at least a dozen other groups, especially Wangkangurru, displaced from their country. At most, there were 150 people in all, with numbers falling due to introduced diseases and a low birth rate, so the 200-seat church was never full.

The Reverend Johann Georg Reuther arrived in December 1888, less than two months since he had landed in Australia from Germany. In between, while visiting South Australia's main Lutheran community in the Barossa Valley, the twenty-seven-year-old Reuther met Pauline Stolz. She was thirty-four years old, a mother of three small children and the widow of another pastor. When Reuther set out for Lake Killalpaninna, eight days after meeting Stolz, they were engaged, and, in February 1889, they married. Lake Killalpaninna was their home for the next seventeen years, during which time Reuther wrote the most extensive account of the place of the long-haired rat in Aboriginal culture, as part of a thirteen-volume ethnography, only possible because of the cooperation and involvement of Aboriginal elders.

Reuther arrived with little English, but improving it was a low priority as the Lutherans used German at all their Australian missions in the late nineteenth century. Reuther's challenge was to learn Diyari. Aided by word lists and grammatical studies prepared by his Lutheran predecessors since the mission's foundation in 1866, Reuther learned fast and was thrilled to deliver his first evening service in Diyari after seven months. Yet he found his work often lonely and difficult and occasionally hazardous. He was at greatest risk when he learned the Diyari planned one of their ceremonies and he sought to disrupt it by delivering a Lutheran service. When the Diyari asked on one of these occasions whether he had a gun, Reuther replied he had. Asked to show it, he drew out a bible.

It was standard practice for Lutheran missionaries to translate Christian texts, but Reuther and his fellow missionary Carl Strehlow were particularly ambitious. In 1893 they set to work on the first translation into an Aboriginal language of the complete New Testament. When Strehlow left in 1894, Reuther persevered alone in between managing the mission, teaching in its school, preaching and much else. In 1895, he recorded: 'Thank God. I finished the translation of the Diyari testament.' Its publication in the Barossa Valley took two years because all the proofs had to be sent to him at Lake Killalpaninna. In 1897, the *Testamenta Marra* was in print.

Reuther's ethnography was even more extraordinary. He embarked on it, he maintained, since 'a missionary without a thorough knowledge of the language and customs of his people, is, in the best instance, like a watch that works but without hands'. Reuther searched through Aboriginal 'legends and the god-and-spirit world of heathendom in an attempt to discover points of contact with the Christian faith and thereby destroy their pagan concepts'. But his manuscript conveys something else—a passion for trying to comprehend and record all he could about the Diyari and other Aboriginal people at the mission. His religious superiors accused him of seeking secular glory as an ethnologist.

The impact of Reuther's work has been slight. While his manuscript was partially translated from German into English in the 1930s, it was not fully translated and published until 1981 and then only on microfiche. Yet it is one of the most remarkable records of Aboriginal language, culture, geography, religion and individual lives. Reuther's manuscript includes a dictionary of more than 4200 words, with long lists of usages and extensive commentary about them, so the entries usually run for many lines and sometimes for several pages. It also includes brief biographies of 300 individuals, and explanations of about 2500 place names also recorded in ink on an accompanying silk map—one of the most compelling European expressions of Aboriginal people's relationship with the land.

As Luise Hercus and Vlad Potezny observed, Reuther 'showed extraordinary esteem and understanding for Aboriginal traditions'. Without Reuther's almost boundless curiosity and appetite for information, 'much knowledge would have been lost'. When he was focusing on Aboriginal placenames, he would work with Aboriginal elders in his study and ask them: '1. What was the meaning of a place-name? 2. Why was it so called? 3. Who named it?' Sometimes they would discuss a word 'for a terrible long time, trying to get to the bottom of it'.

The long-haired rat or *majaru*, as Reuther spelled it, looms large. He experienced at least one plague—most likely during his early years at the mission when a succession of floods arrived down the Cooper. Reuther was so excited by the first, in 1889, that he went out to see the approaching water. 'What a wonderful sight,' he observed. 'How God the Heavenly Father provides for his creatures. Even the desert must bloom.' The next in 1891 was even more spectacular as the water in the Cooper was supplemented by huge local falls. Another flood followed in 1894; yet another in 1906, Reuther's last year at Lake Killalpaninna, when the Cooper stretched ten miles wide and Lake Eyre filled at least partially. Although there were no reports of the rats anywhere in 1889 and 1894, 'millions of the vermin' were said to be coming down the Diamantina and Farrar's Creek in 1891, and there were more accounts of irruptions in Queensland and the Northern Territory in 1895.

Reuther suggested there were two types of *majaru*, differentiated environmentally and physically. One was a sandhills rat with darker fur, the other a plains rat. While one of Reuther's missionary predecessors, Carl Schoknecht, defined *majaru* as the 'wandering rat', Reuther defined it as the 'migratory rat'. The distinction would be significant if Schoknecht and Reuther were recording different Aboriginal understandings of how the rats moved into new terrain. 'Wandering rat' suggests almost objectless movement; 'migratory rat' suggests purposeful travel from one place to another. Most likely, the difference is of no account. Both Schoknecht and Reuther thought the *majaru* was a form of 'Wanderratte'—the German term for *Rattus norvegicus*, reflecting its relatively recent invasion of Europe from Asia. Reuther described the *majaru* as spreading 'locust-like', perhaps the only time these two forms of plague were compared.

The majaru features several times in Reuther's listing of place names. One, which remained long in use—familiar to the senior Wangkangurru man, Mick McLean, in the 1970s—was *Majaru-mithi* or Mira Mitta on the Birdsville Track. This 'rat place' or 'rat camp' was literally 'rat's shining'. Its naming involved Ngurakarlana, one of the *muramuras* or creation ancestors, occasionally human but more often anthropomorphic animals, who shaped and named the landscape in the Diyari's Dreaming. *Majaru-mithi* was where Ngurakarlana observed a multitude of rats with 'lovely glistening fur'.

Another of these place names was *Majaru-woniji*, to the east of Lake Killalpaninna, in the territory of the Yandruwandha on Cooper Creek. *Majaru-woniji* was where the rats began, perhaps the site which the Diyari thought of as their place of origin when they came to Lake Killalpaninna.

Reuther's dictionary reveals more. He recorded that the *majaru* was named after a tree found in Queensland, the *maja*. The rat's permanent habitat—or refuges in times of drought—was where the *maja* grew. Reuther elaborated from his own observation: 'The migratory rat travels down country from Queensland in massive hordes, usually after a sequence of good seasons, and gnaws and devours trees, shrubs and grass. It is eaten by the Aborigines with great relish.'

This account was consistent with the Diyari's understanding that a 'very fruitful land' lay to the east along Cooper Creek which brought the floods that episodically filled Lake Killalpaninna. Rather than following Reuther in thanking God for the 1891 flood, the Diyari 'thanked the blacks in Queensland'. Reuther recorded they sent gifts to two *muramuras*, the Mudlatjilpitjilpi, who were venerable women. If satisfied with the Diyari's gifts, the Mudlatjilpitjilpi sent the water.

Many other entries convey the Diyari's interest in the rats' movement across the country and excitement at their arrival. *Majaru wapana* means rats are on the move or migrating. *Majaru wokarana* is for rats to arrive. *Majaru kunngara* is the sound of the rats coming through the grass in droves. 'The rats are not far away now; can't you already hear their rustling!' is how Reuther evoked it. Both *tapa-tap-ta-teri* and *tapi-tapi-jindri* are the sounds of migratory rats eating. *Tidapirikana* is how the rats ate their way swathelike through the grass.

Wadimajaru is particularly significant. It means the rats were eating the waddi tree, a type of wattle classified by scientists as *Acacia peuce*, renowned for its strong, durable wood and resemblance to casuarinas, now found in only three locations on the edge of the Simpson Desert. One stand is north of Birdsville; a second is south-west of Boulia in Queensland; the third is at Old Andado Station in the Northern Territory. All have been the sites of significant rat plagues.

Reuther identified the *majaru* as one of fifteen rodents eaten by the Diyari—though two were already extinct, at least locally. The one restriction on the Diyari eating the *majaru* resulted from its being a totem. Because the Diyari looked on their totem as family, members of the *majaru* clan did not eat the rat. Otherwise, the *majaru* enriched the Diyari's diet and constituted a great source of fat which the Diyari rubbed on their bodies to keep their skin soft. There is nothing to suggest the *majaru* attacked the Diyari or pillaged their possessions. Reuther's long list of usages of *matana*—to bite—includes the expressions for stings by scorpions, centipedes and wasps and bites by dogs, snakes, mosquitoes, ants, spiders and gnats. Another list for *japa*—to be fearful or wary—similarly includes snakes and ants. The *majaru* does not feature.

The long-haired rat probably loomed large in the cosmology of most if not all Aboriginal groups who encountered it. A little is recorded for the Adnyamathanha people of the Flinders Ranges, but the only substantial record is for the Diyari. The *majaru* appears several times in Reuther's account of the *muramuras*. As historian Philip Jones and anthropologist Peter Sutton suggest, the most important function of the life-stories of these creation ancestors probably related to the control and use of land. One involved a rat *muramura*, Matjamarpinana, who was the focus of Diyari ceremonies for the increase of the *majaru*.

Such 'increase ceremonies', 'increase rituals' or 'increase rites', as they have been dubbed, were staged by many Aboriginal groups in relation to animals that formed an important part of their diet. But as anthropologist John Morton observes, these ceremonies typically seem to have been directed at 'reproducing the country in the broadest sense', not just the maintenance or propagation of desirable species. While Aboriginal women generally hunted the rats, men appear to have played a large part in the rituals involving them. Those who had the rat as their totem would have been the choreographers of these festivals.

A key word was *wonkana*: to sing, chant or enchant. *Wonkana*'s uses, listed by Reuther, included the increase of seeds, wild onions, wood grubs, goannas, frogs and rats. *Majaru wonkana* meant to sing to enchant the rats and increase their numbers. A related term was *kuterina*, with *majarani kuterina* meaning to cast a magic spell over rats or enchant or bewitch them. Because *ngankana* meant to repeat to the deity the invocatory songs of the *muramura* on behalf of game animals, *majaru ngankana* were the songs to produce rats.

These ceremonies involved the shelters which the Diyari called *poonga*, but South Australian colonists called *wurleys* after the word used by the Kaurna people from around Adelaide. As Isabel McBryde has described, they were 'solid beehive-shaped huts built on frames of bent branches' which were 'thatched with reeds of leafy boughs sealed with an insulating layer of sand' providing 'protection against wind and rain'. Reuther recorded that the rat *muramura* erected these shelters when he wanted to increase the *majaru* and the Diyari followed suit, building large shelters 'whenever a sacred ceremony for the increase of the *majaru* rat is performed'. In a rare acknowledgment of continuation of tradition, Reuther admitted that the Diyari erected shelters for the ceremonies to increase the *majaru* 'to this day'.

The Diyari may also have participated in the rat's spread. *Majaru kurakana* meant to free the rats. Reuther noted that to get the rats across a river, an Aboriginal man 'catches a rat, puts it into his bag which is wrapped or hangs around him, carries the rat across the water and releases it on the other side. Within two or three days a large number of these rodents have followed their sister across the water.' He also wrote of this transplanting involving two rats—one male, one female; a father and mother.

'After this happened, the [place] is said to have been literally full of rats within a few days.'

'Enchantment of the rats' also occurred at the end of an irruption which, the Diyari 'generally assumed', was a result of the rats being 'sent home by means of witchcraft by a worshipper of the *majaru muramura*'. If the sorcerer was found out, all that protected him 'from being killed' was that he 'performed the enchantment on behalf of his child, his mother, his brother, or his deceased father'. Enchantment also occurred when a Diyari man died and one of his relatives sought revenge for his death by trying to get the rats to move so 'the people shall have them to eat no more'. The relative did so by secretly taking a rat, placing it in his net bag, feeding it on grass and seeds and moving it, then setting it free, saying, 'You will think of your native habitat; return to your own country! There you will stay; you will summon your friends [the other rats] to come down to you.' Then all the rats retreated.

Majaru tiringana denoted the disappearance of the rats—a time of great stress for the Diyari because of their rapid loss of a great food source. Reuther recorded that the departure of the rats was 'a sure sign that many people are about to die' from what he identified as a type of leprosy or 'warts on the face'. The Diyari thought this leprosy was sent by Mudlatjilpitjilpi, the *muramura* whom they thanked for sending floodwaters down the Cooper. Perhaps this 'leprosy' was a reference to smallpox which may have reach South Australia's north in the 1830s. Reuther thought this sickness was probably due to the Diyari's 'overindulgence' in eating rats. If the sores discharged pus, the sufferer died, according to the Diyari; if the sores bled, he or she lived. The Diyari thought that porcupine or spinifex grass provided 'some sort of immunisation', prompting them to cover the inside and outside of their shelters with it. Still, the Diyari looked forward to the *majaru*'s return.

CHAPTER 16

Villosissimus

No one noticed that the centenary of Europeans in Australia was a turning point for the long-haired rat. After 1888 there were no reports suggesting it extended into Victoria, and its terrain in South Australia contracted sharply, as did its reach into New South Wales. Settlers close to the Darling would not see it. Nor would those back of the Lachlan, long the abode of Kenric Harold Bennett. But scientists would find it where they had not looked before, starting in the Northern Territory. A pair probably found near Tennant Creek prompted Edgar Waite, the head of the Australian Museum's mammalogy department, to revisit the rat's classification as *Mus longipilis* and rename it *Mus villosissimus*. Many more specimens were collected in the Barkly Tableland, which began to emerge as a heartland of the rat, and the most influential tale about it was set in the Northern Territory too.

'Regards to inquiring friends' was how Bennett, still the isolate on Yandembah, concluded a letter to Edward Pierson Ramsay in 1889. Ramsay had become Bennett's agent for Aboriginal implements which Bennett continued to collect—finding private buyers for them, and taking a small cut on the sale price, in addition to Bennett giving him a tomahawk. One of those likely to have inquired after Bennett was Alfred North, a jeweller-turned-ornithologist, initially employed by Ramsay to arrange his private collection of Australian birds' eggs, who then worked under Ramsay at the Australian Museum preparing the first catalogue of its eggs. When he drafted the introduction to this catalogue in 1889, North's most extensive tribute was to Bennett for contributing 'largely to our knowledge'.

This contribution was all the greater because, in one of those typical oscillations of the Australian climate, the intense El Niño of 1888 was followed by a big La Niña in 1889, which Bennett judged one of the best years for collecting since 1870. Aged fifty-four, he climbed a seventy-foot tree and waded and swam into swamps in search of the rare. His greatest find was two nests of the glossy ibis, enabling him to become the first colonist to collect its eggs, and prompting his sixth article in the *Proceedings of the Linnean Society of New South Wales* and, appropriately, his one contribution to the British ornithological journal, *Ibis*. Otherwise, Bennett let North publish his discoveries. When North's catalogue was delayed by printing problems, he added an appendix thick with new material from Bennett. But, like Ramsay, North ignored Bennett's observations of the letter-winged kite—continuing to classify it as diurnal and identifying field mice as its sole food.

The eastern hare-wallaby, a small marsupial with a prodigious leap, classified by John Gould in 1841, had been one of the most common animals on the plains of south-eastern Australia. In late 1889, after receiving a new supply of collecting materials from the Australian Museum, Bennett caught one, preserved it in spirits and forwarded it to Sydney. While he would have known that it had become rare, along with many other small marsupials, Bennett could not know that his specimen would be the final example collected before it became extinct. The museum was similarly unaware, pricing it at just five shillings when it had paid him ten shillings for a bettong and £1 per wallaby. Having long aspired to be honoured for collecting the first of a species of bird—having particular hopes for a form of brown hawk, which he hoped Ramsay would call *Hieracidea bennetti*—Bennett had caught the last of a marsupial.

He also identified a new means of water conservation—a subject that had long engaged him because his uncle William Brodribb had such appetite for land without access to creeks or rivers. When Bennett excavated a tank in a swamp near Moolah's homestead in the late 1870s, he covered it with rushes a foot deep to reduce evaporation. In 1889, Bennett suggested another solution to Frederick Turner of the New South Wales Department of Agriculture involving the aquatic fern, *Azolla rubra*. Having seen this fern spread over the entire surface of lagoons, Bennett proposed that it be placed in water tanks to limit if not stop evaporation.

Bennett would have liked to stay on Yandembah, but with it restocked and carrying 16,500 sheep, his seventy-eight-year-old mother decided to sell it and quit the land. All the family agreed apart from Kenric, who feared he would be 'amongst the great unemployed once more', little trusting a 'promise' of work from Ramsay involving his private bird collection. Expecting Yandembah would sell, which it did not, Bennett moved in 1890 to Sydney, an alien environment for such a man of the outback. Although Ramsay gave him no work, Bennett chose a house in suburban Ashfield because it was close to Ramsay's home. When he visited the Australian Museum, he would have taken pride in its display of material he had collected, including a case largely devoted to Aboriginal implements from between the Lachlan and the Darling. His last gift to the museum in March 1891 was part of an Aboriginal skull; there is no record of where or how Bennett obtained it. On walks with Alfred North in the local bush, they looked for birds and eggs. Once Bennett climbed a tree for eggs. Once North climbed on his shoulders to do the same. That June, Bennett had a stroke and died, aged fifty-six.

His last paper, long in the research and writing, appeared that March in *Records of the Australian Museum*, which Ramsay had just founded. Sydney's *Evening News* considered these *Records* 'of little interest to the general public'. The title of Bennett's paper, 'Notes on the Disappearance—Total or Partial—of Certain Species of Birds in the Lower Lachlan District', was consistent with that. But for all it was a regional study, confined to part of New South Wales, the weekly *Australasian* in Melbourne republished it that April, recognising its larger interest. While identifying him as 'K. H. Bennett F.L.S.' to demonstrate his authority as a Fellow of the Linnean Society, the *Australasian* retitled his piece 'Bush Birds Missing' to attract the readership it deserved.

Bennett primarily blamed the settlers' livestock. He knew, from long experience, how sheep and cattle ate out saltbush and native grasses, often leaving the land bare. Bennett also blamed the 'prevalence of the domestic cat (gone wild)'—perhaps the first colonist to recognise its destructiveness. The first species that Bennett discussed was the flock bronzewing, so abundant when he first arrived in the back country. Less than thirty years later, it had 'entirely vanished'—part of its much larger decline, which more recent ornithologists have likened to the demise of the passenger pigeon in North America. Ten more species followed before Bennett identified three others, originally 'few in numbers', that had benefitted from colonisation and 'become plentiful and permanent'.

The Linnean Society's president, William Haswell, a graduate of Edinburgh, recently appointed as Sydney University's inaugural Challis professor of biology, acknowledged Bennett's death in a brief, patronising paragraph in his annual address. Haswell identified Bennett as 'one of that not too numerous school of educated bush-naturalists, who spending their lives in the country, engaged in pastoral and other pursuits' was 'yet sufficiently in touch' with societies such as the Linnean 'to permit of their observations being recorded and utilised'. Bennett had contributed 'a number of new facts...chiefly about birds—their habits, their nesting, and their eggs' and 'at one time devoted much attention to native weapons, implements, and utensils, in which he was well versed'.

Botanists continued to rely on his work. As Frederick Turner initiated trials of the use of *Azolla rubra* in water tanks to reduce evaporation, he invoked Bennett's observations. In *The Forage of Plants of New South Wales* published late in 1891, Turner acknowledged his indebtedness to Bennett for several specimens. J. H. Maiden, who had begun working for the New South Wales forestry department as well as curating the Technological Museum, displayed most regard for Bennett in an article in the *Sydney Mail* in 1892 about 'Australian Economic Plants'. Maiden lauded Bennett as a 'perfect mine of information about the natural history of the back blocks…whose recent death all students of Australian natural history deplore'.

A few years later, Bennett's brief employment as a collector by the Australian Museum in 1883 bore belated fruit. Alfred North examined a group of small crows collected by Bennett, identified them as a new species and named them *Corvus bennetti* because Bennett had collected this bird, recognised its distinctive size, appearance and song, and was an 'esteemed friend' who contributed 'so largely towards completing a knowledge of the Australian avifauna'. But while Bennett had loved the hawks which he thought might become *Hieracidea bennetti*, crows were the only birds that he abhorred, enraged by how they frustrated his attempts to collect the nests and eggs of other species. In a letter to Ramsay, he recorded taking—and keeping—'a vow to spare neither age nor sex' of these 'infernal birds'.

In 1892, John Gould's account of the rat finally reached a broader Australian audience when the Australian Museum published a *Catalogue of Australian Mammals* by one of its former curators, James Ogilby. This *Catalogue* was a counterpart to Gould's *Handbook of Australian Birds*, providing basic descriptions of more than 200 mammals including the long-haired rat. It was a book Bennett would have loved—and could have afforded to buy—intended to enable the bush naturalist 'round his camp fire in the evening, to determine the specimens which he has obtained during the day'.

Gould's blunder in naming the rat *Mus longipilis* was exposed soon after due to a wave of collecting in central Australia triggered by a visit in 1894 by Baldwin Spencer, the professor of biology at the University of Melbourne. 'All the people…who know their alphabet now seem to be going in for eggs, rats, bugs and shells,' remarked Ernest Cowle, a South Australian policeman stationed at Illamurta Springs. They did so by employing teams of Aboriginal people whom they referred to as 'Niggers' or 'Nigs'—commonplace colonial terms expressing the settlers' deep-seated racism. Older Aboriginal women, the traditional gatherers of small game for food, were the best collectors of 'rats', which continued to be the settlers' general term for small mammals.

Frank Gillen, who was the prime administrator of central Australia as the Alice Springs postmaster, magistrate and protector of Aborigines, was one of Spencer's collaborators. The results initially were disappointing. In September 1894, Gillen reported: 'No rats since you left though Nigs have been on the look out, but we have just had some nice rain, the weather is becoming warm and we hope to have some before next mail.' A month later: 'The Niggers captured two rats...but unfortunately both got away.' Then: 'So far we have not been able to get a single rat...although I have fitted out several expeditions.' And: 'Still unable to procure more rats though niggers are always on the look out.'

A visit to Tennant Creek in 1896 finally saw Gillen full of confidence. 'If you don't find something new or at least very rare in the collection,' Gillen wrote, 'strike me off your list of contributors.' After Spencer received this collection in Melbourne, he had the rodents identified at the Australian Museum by Edgar Waite. Such work was generally frustrating for colonial scientists because of the dearth of specimens and reference books in Australia and 'the multiplication of species and the number of species insufficiently described', especially by John Gould. But consideration of *Mus longipilis* was simpler since its type specimen, collected by Thomas Wall and classified by Gould, was in Sydney where it had been stuffed by a taxidermist in the pose of an English stoat with a narrow waist and arched back, mounted for public display and faded through long exposure to light. When Waite examined the rats sent by Gillen, he found that two were the same species.

'It is gratifying to find that this rat has again been obtained,' Waite wrote in a paper presented to the Royal Society of Victoria in 1897. One, female, would go to the South Australian Museum in Adelaide so it would have an example of the rat. The other, male, stayed in Sydney in the Australian Museum—labelled incorrectly as dating from 1894, but identified perhaps accurately as coming from Wycliffe Creek, north of Tennant Creek. Since both Tennant and Wycliffe creeks were far from the rats' normal range, offering them little suitable habitat, they were probably there as a result of an irruption sketchily recorded in western Queensland and the Barkly Tableland of the Northern Territory in 1895.

Waite recognised that the rat needed renaming because Gould had used *Mus longipilis* when the Zoological Society of London's curator George Waterhouse attached this name to one of Charles Darwin's South American specimens. Since Waite otherwise agreed with Gould's name—considering the rat's hair its 'most striking feature'—he called it the hairy rat *Mus villosissimus*. He also described it with unprecedented precision, providing many measurements of external parts of its body and describing its skull and teeth in particular detail, because taxonomists had realised these characteristics provided one of the best means of distinguishing rodents.

Australia was entering one of its most severe droughts. This 'Federation Drought', as it has been dubbed, since it took place at the time of the creation of the new nation, was a tipping point for many native species, resulting in a sharp drop in biodiversity. Because of the general dearth of rain, there were no big irruptions of the rat, but it still spread across small areas in the wake of good local falls. A report of a trip down the Flinders and Saxby rivers in 1900 included an account of 'a plague of...the large grey rat' which 'played all over' the travellers at night.

Another irruption occurred in central Australia a few years later when the drought was over. As recalled by a resident of Alice Springs, a few rats appeared there in September 1904, then many more, travelling 'due north to south' and 'clearing up everything as they went' for four weeks, before their numbers dropped. As usual, 'small vegetable gardens disappeared in quick time. They then started on any kind of leather, such as harnesses and saddles, and on clothing, and were particularly keen on anything with a little grease on it.' This settler remembered—or misremembered—them as 'half the size of the ordinary everyday rat', with a bushy tail.

The long-haired rat was collected for the first time in big numbers due to the cachet of London. Oldfield Thomas, a zoologist at the British Museum, was the catalyst. When he told John Forrest, the managing director of the North Australian Pastoral Company, that no satisfactory collection had been made of mammals of the Northern Territory and of northern Queensland, Forrest and another of the company's owners Sir William Ingram hired a collector to secure material for the museum. Their choice was a young Englishman, Wilfred Stalker, who had worked in New Guinea and would die there on another collecting expedition. For two years, Stalker worked on Alroy Downs and Alexandria Station, two giant cattle runs in the Barkly Tableland. While primarily covered with Mitchell and Flinders grasses, they included desert, pebbly ridges, black soil downs and coolibah-lined rivers, enabling Stalker to collect in many different habitats.

This part of the country had been devastated by the Federation

Drought but began recovering after exceptional rains in March 1903 and, by mid-1904, was in 'magnificent' condition with 'wavy green grass stretching as far as the eye can see' and peabush not just ten-foot high as in 1870 but fifteen-foot and very thick. When Stalker arrived late in 1904, the conditions were deteriorating, and they worsened again in 1905 due to the failure of the tropical wet season. On the whole, Stalker could catch little. The exception was a long grey rat, which he took to be *Mus villosissimus*. It was especially abundant about the station at Alroy Downs where it was a 'serious pest'. That winter, as the drought continued, this rat was almost all Stalker could catch, even with Aboriginal help. It remained 'quite a problem' at the station and sprang all Stalker's small traps. Then its numbers plummeted, which Stalker attributed to the continuing drought, and all he could find in late 1905 were the skins of dead rats in old burrows.

The 'bush-travelling rat', as a journalist with the *North Queensland Register* dubbed it, was also abundant elsewhere. A plague was reported not only on the Flinders River but on Wave Hill station on the Victoria River—the river that John Gould had mistakenly given as the source of Thomas Wall's rat from 1847. After Stalker sent twenty specimens to the British Museum, Oldfield Thomas confirmed they were *Mus villosissimus*. Thomas also recognised their similarity to the pair of rats from the Liverpool Plains sold by Gould to the British Museum in 1841, which John Edward Gray had classified in 1843 as *Pseudomys greyi*, another name already used. Because Thomas thought the coats of the Liverpool Plains specimens were exceptionally long and thick, he defined them as a sub-species, though more recent scientists have rejected this distinction.

'Rat' had long been a term for self-interested abandonment

and betrayal—linked to Pliny the Elder's observation about mice deserting a building set to collapse and Shakespeare's line about rats quitting a sinking boat. In Australia it began to develop a distinctive usage. A 'Labor rat' was a traitor to those on the left of politics, 'unspeakable, contemptible', who 'should receive the undying hostility of the working class'. The way in which rats were perceived was also transformed by the French physician Paul-Louis Simond who, as bubonic plague spread around the world in the 1890s, concluded that *Xenopsylla cheopis*, a flea that usually lives on *Rattus rattus*, transmitted the plague when its original rodent host died and it switched to humans.

When Simond's thesis was accepted with particular speed in Australia after the plague reached Sydney in 1900, rats became the stuff of an unprecedented 'war of extermination' which, along with a new vaccine, kept the number of people to die of plague that year in Sydney to just 100. The New South Wales Government and Sydney Council distributed poison free of charge, fumigated sewers, employed professional rat catchers, encouraged members of the public to join in the hunt through a twopenny rat bounty, soon raised to sixpence, and employed a special rat incinerator. The number of rats killed in Sydney in the first three months was more than 45,000; in eight months, over 100,000. A typical editorial identified rats as 'scrofulous', 'retrograde' and 'a professional parasite'. Yet it was not just that rats were identified as pivotal to the spread of this 'Black Death' as the pandemic was widely dubbed, echoing that which started in the fourteenth century. Rats were also blamed for the first Black Death, as they generally still are.

Heber Longman of the Queensland Museum pronounced: 'The rat is one of the serious problems with which modern civilisation has to deal', tracing human horror of it to 'the beginnings of historical time'. Black and brown rats were his prime concern. He acknowledged that other species posed less of a threat, but declared that 'no kind can be regarded as harmless or a desirable feature of a civilised environment'. Yet there were—and continue to be—doubts about the rat's part in the first Black Death, partly due to the dearth of contemporary reports of rats dying. As one of Australia's leading public health officials, J. S. C. Elkington, put it in 1903, 'The study of past epidemics in Europe shows scarcely any reference to coincident mortality amongst these animals, and it is scarcely possible to believe that such a striking coincidence would have escaped mention by the numerous, and often acute, contemporaneous observers.'

The long-haired rat reached its greatest audience through London's *Wide World Magazine* which first appeared in April 1898 under the motto 'Truth is Stranger than Fiction'. Its rat story was part of 'The Adventures of Louis de Rougemont' who claimed to have spent thirty years living with Aborigines in Australia's far north-west. When the *Wide World Magazine* published the first instalment in August 1898 as a 'contribution to science' confirmed by experts as 'perfectly accurate', demand was so great that the magazine printed 100,000 extra copies of its next issue featuring the second instalment. This interest grew again when de Rougemont talked at a meeting of the British Association for the Advancement of Science and enjoyed surprising success in fielding questions about his tales of flying wombats and riding turtles in the water.

Then London's *Daily Chronicle* revealed that 'de Rougemont' was Henri Louis Grien or Grin, a 'hoaxer', 'fraud' and 'modern Munchausen'. While he had worked on a pearl lugger in the north during more than twenty years in Australia, he had never lived with Aborigines and had constructed much of his story in the reading room of the British Museum in London. Yet, having caught the public imagination, 'The Adventures of Louis de Rougemont' retained its audience, with its later instalments maintaining the *Wide World Magazine*'s vast circulation, the British book of the serialised parts selling 50,000 copies and American and foreign language editions soon following.

Grin, always better known as de Rougemont, recognised the wondrous nature of the rat's irruptions in the magazine's February 1899 issue. 'I question whether a spectacle so fantastic and awe-inspiring was ever dealt with even in the pages of quasi-science fiction,' he wrote. He also understood that Aboriginal women 'were responsible for the catching of the rats, the method usually adopted being to poke in their holes with sticks and then kill them as they rushed out'. But most of what he wrote was fiction. He cast the rats as diurnal, so that when he first saw an irruption one morning 'the whole country in the far distance was covered with a black mantle…of living creatures'. He identified the rats as all-consuming, blaming them 'for the horrible deaths of many native children'. One of the accompanying illustrations by Alfred Pearse, best known for his work for the *Boy's Own Paper*, was the first to show the rats en masse. Relying on de Rougemont's account, Pearse pictured how 'snakes, lizards—aye, even the biggest kangaroos—succumbed after an ineffectual struggle', and de Rougemont and his 'ever-faithful' dog, Bruno, survived only because his Aboriginal 'wife' instructed him to take refuge up a tree.

In the wake of these 'Adventures', some of the first squatters to take up land in the Channel Country in the early 1870s recalled that they too had encountered prodigious numbers of rats travelling in daytime. Most likely, such stories had long been rife, fueling de Rougemont's account, but perhaps the squatters were influenced by his tale. John Conrick, who established the Nappa Merrie Station on Cooper Creek, claimed to have seen 'thousands of rats' while 'riding across the plains' as there was 'a regular plague of them at the time'. Tom Archer, who settled on the Thompson River, recalled meeting 'millions of rats travelling south along pads they had formed' following the same route 'similar to sheep pads'. It took the party 'about 48 hours in passing them—say 20 miles for two days' travel with the cattle'.

A very different account came from Sidney Pearson, who traversed the continent towards the end of the Federation Drought, heading north to prospect for minerals, then returning south with cattle. In a remarkable essay wasted on a small audience in regional Victoria in 1904, the nineteen-year-old Pearson displayed great understanding of the heartland of the long-haired rat and its irruptions. He observed: 'On the spinifex deserts and great Mitchell grass plains of the Northern Territory is the home of the grey bush rat—the same rat that in years of plenty migrates toward the coastal districts, and within the knowledge of old hands has been seen in millions on the dividing range in Queensland, the Lachlan and the Murrumbidgee.'

CHAPTER 17

Three Cheers for the Diamantina

The first daily log of the rat was an accident. When the naturalist Sidney Jackson set out for western Queensland in 1918, he was seeking the letter-winged kite. Having encountered it just once—on the north coast of New South Wales, where he found a pair and secured a clutch of their eggs—he hoped to find one of the kite's main breeding grounds so he could secure male, female and juvenile specimens, more eggs, and study the kite's habits and habitat. He was oblivious to the kite's links with the long-haired rat until he reached the Diamantina River. 'I never heard of such a pest as these brutes and did not expect to meet with them here,' Jackson recorded in his diary.

The architect of his journey was H.L. White, one of Australia's wealthiest pastoralists, who became obsessed with the letter-winged kite when he read John Gould's *Handbook of Australian Birds* and learned how Samuel White—no relation—had encountered this kite on the Cooper in the early 1860s. As H.L. White used his fortune to build Australia's best collection of bird eggs, he always instructed his collectors to look for the letter-winged kite, assuming almost nothing was known about it. White's specialty, as nature writer Alec Chisolm observed, was pursuing 'the little known birds of Australia's out of the way places'. The 'romance' of the kite was due to 'the semi-mystery attaching to its nesting places'. Had Kenric Harold Bennett's observations gained the recognition they deserved, White would not have been so interested. But having been ignored by Bennett's scientific contemporaries such as Edward Pierson Ramsay and Alfred North, they continued to be overlooked.

White's hopes of discovering more came to fix on Davenport Downs—a vast cattle station on the Diamantina River in western Queensland, where the kites must have often nested, as this part of the Diamantina was a prime domain of the long-haired rat. White's hopes rose in 1914 when he learned that the owner of Davenport Downs had seen one of the kites' breeding places. When White pursued this lead from his home at Belltrees in the Hunter Valley of New South Wales, the station's manager had nothing to report. When White wrote again in 1916, the bird 'appeared to be there, and was apparently of nocturnal habits'. When White asked in 1917 for a skin, he received one of a letter-winged kite, but a set of eggs that followed were those of the closely related black-shouldered kite, which occasionally hunts on moonlit nights but is otherwise diurnal, primarily hunting in the early morning or late afternoon.

Sidney Jackson had become obsessed with eggs as a boy in the 1880s and assembled a spectacular collection—aided by a tree-climbing ladder more than a hundred-foot long, developed and patented by Jackson himself, which was raised by shooting a fine line, weighted with lead, from a catapult, over a limb of the required height. After selling his egg collection to White in 1907, Jackson secured just the kind of employment Kenric Harold Bennett had desired. Jackson became White's curator, and went on many expeditions to places that offered 'a virgin field', to get 'rare stuff'. Although Jackson did not look the part—White considered his appearance was 'between that of an actor and an Italian count'—White reckoned him the best egg collector in Australia.

World War I was curtailing public science as governments slashed funding. The Australian Museum was forced in 1915 'to cease purchasing specimens and books, discontinue publications, and cancel the winter lectures'. But the rich could pursue what they wanted. White also had a public purpose since he intended to donate his egg collection to the National Museum of Victoria. To send Jackson to Davenport Downs, White teamed up with his fellow pastoralist and avid bird and egg collector, J. H. Bettington. Because Davenport Downs was 260 miles from the nearest railway station, with communications beyond it 'very uncertain', White and Bettington decided Jackson should make the 1400-mile journey from the Hunter Valley by car. Bettington provided a new Australian-made Buick with a chauffeur, Harry Cottrill, who was another adept tree-climber but much lighter and able to collect eggs from branches too thin to take Jackson's weight. White and Bettington also had supplies and equipment for Jackson and Cottrill shipped to Townsville, then transported by rail to Winton.

Jackson's one other war-time collecting trip, to Tasmania, had been a failure, so he was unusually anxious when he set out with

Cottrill from the Hunter Valley in June 1918. Eighteen days later, 'travelling by easy stages without as much as a tyre puncture', they reached Winton, where Jackson was relieved to find an old Aboriginal man who had seen many letter-winged kites and knew they nested in coolibah trees, their usual roosting places along inland watercourses. The remaining drive to Davenport Downs was through country birdless except for the Australian raven and *Corvus bennetti*, the crow named after Kenric Harold Bennett. On reaching Davenport Downs, Jackson immediately saw a few letter-winged kites in coolibahs on a waterhole which, Jackson observed, 'caused us much excitement and pleasure and set our minds at ease'.

When Jackson and Cottrill went on a few miles and discovered a colony in a stand of coolibahs on another waterhole, they camped there and immediately set about tree climbing, to have eggs in hand to justify their journey. Jackson was first, reaching a nest sixteen feet above the ground, which contained three eggs. In other nests, eleven to nineteen feet high, he found four, five and six eggs in each. But because some nests were up to thirty-five feet from the ground, and near the ends of branches leaning well out over the waterhole, Jackson left them to Cottrill who sometimes roped one branch to another since the smaller branches could 'snap like a carrot', especially when it was windy.

The station was in the country of the Maiawali and Karuwali people who called the kite *gidga-gidga*. Jackson did not record their name for the rat but observed they 'greatly enjoyed' a meal of 'large fat' ones, while Cottrill and he delighted in 'very fine beef' provided by the station. They found that the kites' nests were lined with masses of rat fur, and the ground below was thick with pellets of fur vomited up by the birds after eating the rats.

Jackson frequently found thirty to fifty pellets under a nest, once more than seventy. Like Bennett, he never saw the kites hunt during the day.

He reported that they 'roosted all day in the coolibahs, and went out over the plains rat-catching as soon as it was dusk, and kept going backwards and forwards all hours of the night and early morning'. While he 'frequently heard the males calling on returning from the plains with rats on dark nights, and feeding the females at the nests in the usual way', Jackson recognised the kites preferred to hunt by moonlight: 'As a rule, the noise usually started with renewed vigour just as the moon rose, and it did not matter if it was 8 p.m. or 2 a.m.; then the babble of cries and rat-catching proceeded.' Jackson also recognised that, in a striking instance of evolutionary adaptation, the kite 'had become nocturnal, owing to the habits of its prey'. In ornithological and oological terms, the trip was a fabulous success. When Jackson and Cottrill left that August, they gave three cheers for their camp.

Davenport Downs was one of the long-haired rat's heartlands. Frank Jaines, a bushman with a keen interest in natural history, who spent much of the 1880s between the western plains of Queensland, the open downs of the Northern Territory and Lake Eyre in South Australia, encountered the rats there during their great irruption. He later recalled them being 'a terrible nuisance' for about five months at three different camps on Monkira and Davenport Downs cattle stations. 'We had to make pits, and with a balanced stick got 400 and 500 in one night, and had a heap of dead rats which had to be burned.'

The rat was probably there again during its next major plague in western Queensland, triggered by a sequence of La Niña years. In late 1908 the rat 'swept over' Winton for a few days, though with little effect because of the town's 'superabundance of ownerless cats'. While thought to be 'migrating west', it was next reported to the east between Muttaburra and Aramac early in 1909 and again outside Winton. In the middle of the year, it was 'fairly plentiful' in Longreach and in 'myriads' around it. At the start of 1910, a plague outside Hughenden was followed, as settlers had come to expect, by a plague of cats. 'Measly-looking, starved creatures', they died in vast numbers apparently 'suffering from some disease'.

The rat's appearance in 1918 occurred in the wake of another La Niña, which saw much of Davenport Downs under water for the first three months of the year. As reports of rodents soon proliferated, some were of 'millions of mice'. Others were of 'a remarkable plague of rats', so 'anyone driving along the roads runs over numbers of them'. A vegetable garden at Gilliat in Queensland's Gulf Country run by an 'old Chinaman' and 'greatly appreciated by the surrounding residents', was pillaged in a few nights. Sheep 'with holes eaten into their loins, over the kidneys, by rats' were found still alive 'only to die soon after'. Carpet snakes said to be almost eighteen-feet long, an otherwise unrecorded size, feasted on the rodents. In keeping with a novel empirical study by a Boston insurance company that rats nibbled all styles of matches and set fire to all except safety ones, many Queenslanders continued to blame the rats for bushfires. The Flinders and Mackinlay councils called on the state government to ban the sale of wax matches. At the end of the plague, cats abounded, 'devouring carcasses of stray dead cattle, and incidentally all the bird life in their tracks'.

Jackson had been attacked on previous expeditions by ticks, leeches, Barcoo fleas, red-back spiders, March flies and black snakes. He considered the rats much more of a menace. Their apparent lack of fear shocked him as they sat on his boots and started eating them while he wrote in his diary. The surrounding landscape was also frightening. When Jackson heard a gnawing sound near his camp and went to discover its cause, he found about a hundred rats feasting on a dead bullock. Some were eating its hide, others the dried, hard sinews on its bones, which 'they had eaten and combed out like masses of cotton threads'.

The rats' attacks gave Jackson and Cottrill little opportunity to relax as they sat around their fire at night. If they put down anything small such as a pipe, the rats were off with it at once. They even chewed the wooden handles of bush knives and the base of Jackson's lantern. The only containers of food that could be left on the ground without special protection were unopened tins. Jackson and Cottrill suspended the rest of their food from wires and ropes tied to the branches of coolibah trees. While the rats generally could not reach them, one morning Jackson awoke to find the 'fearful' rats had got his damper from what he still called 'a safe place'.

Jackson and Cottrill soon learned the rats could do worse. While the local postman was asleep while camping out, the rats bit his fingers so they bled severely. Jackson wondered why he escaped such assaults when the rats repeatedly ran over him while he was in his swag. But he was more concerned about J.H. Billington's new Buick. Having got it to the Diamantina intact, Jackson did not want to see the rats damage it. He recorded: 'Fortunately they did not get in the motor car, the mud guards preventing them from getting up...they had a great many tries at the rubber tyres, but found their nibbling useless.'

Jackson and Cottrill began by killing as many rats as they could, using some of the corpses as food for young letter-winged kites, which they took from nests and raised by hand until they killed them as specimens. Jackson and Cottrill also filled an old bag with rats' intestines as a lure to catch fish in the waterhole. But Jackson gradually came to accept the rats. When Cottrill took the Buick to collect Billington and his son, who had travelled north by rail to join them, Jackson recorded: 'Lonely again in camp tonight, the rats are my only company.'

A few years before, H. L. White had ordered Jackson not 'to go messing about after snails, beetles, grubs, botanical specimens or any such vermin', suggesting that Jackson's larger interest in natural history inclined him to do so. On another expedition, White instructed: 'I don't want botanical specimens. Stick to the birds and eggs.' Jackson ignored this instruction while he was on the Diamantina where he wrote about the rats almost daily. He also photographed and collected them, amazed by their size and numbers, especially since Davenport Downs was dry, with food for the rats diminishing.

When Jackson first measured some in early July, he recorded they were fourteen inches from head to tail, 'the body alone measuring 8 inches'. Later, he found one seventeen-and-a-half inches from head to tail and nine-and-a-half inches in the body, the biggest recorded. He figured there must have been 'tens of thousands' on the plains at night since 'every inch of ground for miles...was marked with their foot prints'. The rats destroyed vast areas of a yellow-topped plant that had sprung up following floods that February, reducing it to 'fine chaff'. They similarly took to the fluffy flowers of a creeper that grew thickly on the plains and claypans, so the ground was white with tiny pieces, 'like millions of tiny hail stones'. With manifest amazement—and delight—he observed: 'Walking over the plains and the traces of rats everywhere was wonderful. It is beyond description.'

Jackson, who came equipped with camera and tripod—by then standard ornithological equipment—was first to photograph the kite and the rat. But because these animals were nocturnal, his photographs missed much. He did not record the kite's flight or the rats swarming at night on the plains or in his camp. His photographs, all taken during the daytime, showed the kite's young—occasionally still in their nests, more usually ones that Jackson raised for a few weeks at most before killing them as specimens. The rats featured as the birds' food. In one photograph, seven dead rats were carefully lined up in front of three kite chicks, but not yet ready for the birds as Jackson and Cottrill 'always removed the skin of the rat before cutting it up to feed them'.

John Cleland, a pathologist at the forefront of studying bubonic plague, had just brought together many published and unpublished accounts of rat and mice irruptions in Australia. When Jackson wrote about his expedition to the Diamantina for the journal of the Royal Ornithological Society of Australia, he drew on Cleland's research but reached a different conclusion about the main species involved. Cleland assumed that, since Edward Pierson Ramsay's *Mus tompsonii* had been identified as *Rattus rattus*, the rats which Kenrick Harold Bennett encountered in 1887 were too, along with those in other irruptions elsewhere. Heber Longman of the Queensland Museum, who had also recently written about Australia's rat plagues, agreed they were an introduced species but was unsure whether they were *Rattus rattus* or *Rattus norvegicus*. Having brought several bottled specimens to Sydney where he had them examined at the Australian Museum, Jackson announced most were the native long-haired rat.

The impact of Jackson's work was patent in 1926 in *The Wild Animals of Australasia*, the first generalist book about Australian mammals to reach a big audience. 'Only one native rat, as far as we know, exhibits the swarming propensities so well known in this order throughout the world,' observed the first director of Sydney's Taronga Zoo, Albert Sherbourne Le Souef, and the naturalist Harry Burrell. As *Mus* had become a term used only for the introduced mouse, Le Souef and Burrell called this native rat *Rattus villosissimus*. Seventy-nine years after Edmund Kennedy's expedition encountered it on Cooper Creek, the long-haired rat was, at last, identified as Australia's prime irruptive rodent.

CHAPTER 18

The Summer Vacationist

Hedley Herbert Finlayson was a student at the University of Adelaide in 1913 when a disastrous chemistry experiment resulted in him losing his left hand, part of his right thumb, some of his hearing and much of the sight in his right eye. Rather than continue his degree, Finlayson became a demonstrator in the chemistry department, a poorly paid post with no status, and he turned to zoology. In 1931, aged thirty-six, having begun publishing in the journal of the Royal Society of South Australia and enriching the collection of the South Australian Museum, Finlayson became its unpaid Honorary Curator of Mammals, a position created for him. In an era of increasing professionalisation of science and ever-greater deference to academic credentials, Finlayson was an oddity: an autodidact, with a marginal role in urban institutions, whose flair and appetite for fieldwork saw him eclipse his salaried counterparts.

Finlayson transformed understanding of many species, including the long-haired rat, by undertaking a series of four self-funded summer expeditions between 1931 and 1935 to investigate the mammals of Australia's interior. One of his champions was Frederic Wood Jones, Professor of Anatomy at the University of Melbourne, who himself had done exceptional research in the 1920s for his three-volume *Mammals of South Australia*. In 1935, when Finlayson had published only a fraction of what he had discovered on his expeditions, Wood Jones declared: 'No living man has done so much in rescuing from oblivion those sparse but interesting mammals that still inhabit Central Australia, as has Finlayson.' More recently, Steve Morton, one of Australia's foremost ecologists, observed that 'without Finlayson's extraordinary voluntary efforts, we would know little today of what life used to be like for the mammals of inland Australia'.

As Finlayson himself acknowledged, some of his most celebrated work depended on Louis von roon Reese—usually known as Lou Reese—the owner of Minnie Downs, a South Australian station more than 200 miles downstream from Davenport Downs where Sidney Jackson had encountered the letter-winged kite and the long-haired rat. From the mid-1920s, Reese's love of natural history saw him contribute regular bird notes to the *South Australian Ornithologist*, co-author an extended paper published by the Royal Society of South Australia about the flora of the state's north-east and a short, supplementary one about the flora of Minnie Downs. After meeting Finlayson in Adelaide in 1929, Reese collaborated with Finlayson on a survey of the Diamantina, sending him 'a lot of good mammal material'. A few years later, Finlayson paid tribute to Reese for his 'constant contributions' to zoology, ornithology, entomology, botany and anthropology. While many people were ready to undertake museum collecting 'on a strictly cash basis', Reese devoted his own money, time and energy to 'systematic collection of data'.

Finlayson also depended on a Wangkangurru man, Jimmy Naylon Arpilindika, who was born in the Simpson Desert around 1880. As Philip Jones described him, Jimmy Naylon was 'among the last Aboriginal people living beyond the bounds of European influence', remaining in the Simpson with other Wangkangurru until the Federation Drought. Despite the drought's intensity, they could have stayed there, able to find reliable water in the soaks that they called *mikiri*. But, as Luise Hercus explained, 'tales of plenty' spurred them to leave.

When Finlayson arrived at Minnie Downs, he wrote of Reese 'ruling' these Wangkangurru 'patriarchally, and with much contemptuous understanding'. Philip Jones identified the Wangkangurru as 'a captive labour force', paid in rations, but who remembered Reese 'with respect, not simply as a taskmaster'. As Luise Hercus and Peter Clarke recorded, Jimmy Naylon 'was always homesick for his own country', spoke about it 'to his children and grandchildren, and desperately wanted them to see it' and 'shortly before his death in 1965...took his favourite young grandson Jimmy Crombie from Birdsville where he and his family were living'.

According to Finlayson, Naylon's years in the Simpson, 'a real desert', had developed in him 'amazing skills as a hunter'. He was Reese's 'rat boss' and 'mammal expert', who specialised in small animals, which traditionally were hunted primarily by women. Finlayson, who photographed Naylon as well as wrote about him, lauded him as his 'mainstay', the 'king pin', when he first visited South Australia's far north in 1931 in pursuit of the desert or plains rat-kangaroo—the hunt that transformed Finlayson's professional and public identity, though Adelaide's *News* still considered his mammal work a 'hobby'.

John Gould had classified the desert rat-kangaroo in 1843 in the *Proceedings of the Zoological Society of London* on the basis of three specimens in the British Museum—identifying this small, largely nocturnal marsupial as *Bettongia campestris* and reporting that it was to be found on 'the stony and sandy plains of the interior of South Australia partially clothed with scrub'. Gould also illustrated it in his *Mammals of Australia* and Oldfield Thomas of the British Museum gave it a new designation in 1888 as *Caloprymnus campestris* after re-examining the museum's specimens. But apart from one possible sighting in 1878, there were no records of it. When Frederic Wood Jones considered its fate in the 1920s in his *Mammals of South Australia*, he identified this rat-kangaroo as 'one of South Australia's most remarkable animals' and suggested it had 'probably passed away comparatively recently without any trace of it having been preserved, not in the Museum of its own State, but in any scientific institution in Australia'.

Lou Reese enabled Finlayson to rediscover this animal, which Aboriginal people on the Diamantina called the oolacunta. Having seen perhaps twenty over thirty-five years, Reese told Finlayson about it in 1930. Most likely, since he was familiar with Jones's *Mammals*, Reese had at least some sense of the significance of his observations. After Finlayson recognised that Reese's account fitted *Caloprymnus campestris*, Reese sent him its skin and skull, enabling Finlayson to announce its survival in September 1931. When Finlayson decided to head north that December, Reese identified the best area to seek the oolacunta was on Clifton Hills, a neighbouring station on the Diamantina. Reese advised that trapping would fail and shooting would be too damaging. Instead, they should pursue it on horseback with four Aboriginal men, including Jimmy Naylon Arpilindika.

The conventional scientific appetite for bones shaped this hunt. Finlayson's prime goal, he announced before setting out, was 'a revision of the characters of the animal', which required an array of specimens, male and female, of different ages, allowing a much more precise physical description of it. Far from being concerned at killing a creature not collected for ninety years and thought to be extinct, Finlayson considered the oolacunta not at particular risk. While recognising its numbers were generally small, he attributed its apparent disappearance 'rather to lack of systematic collecting than to any sudden change in the status of the animal'. He thought of the oolacunta as 'waxing and waning with varying seasons', with 'a wide but sporadic distribution in the great areas of sand hill and gibber plain country between the Diamantina and Cooper Rivers'. His ambition was to 'get all we could of the oolacunta'—a goal embraced by Frederic Wood Jones, whose first question was, 'did you get many?'

The first—a big female oolacunta and her big joey—were shot by Reese when Finlayson, Jimmy Naylon and he went for a walk on their second afternoon at Clifton Hills. Finlayson 'felt tremendously gratified', he recorded in his diary, 'the success of the trip assured'. But he omitted this episode from his published accounts. Instead, he wrote as if Reese, the four Aborigines and he first caught the oolacunta after pursuing it on horseback because that was much more exciting.

As they rode, the one-handed Finlayson always had his camera at the ready. In the course of one chase, Finlayson even succeeded in making 'two fast exposures on the move', resulting in a breath-taking photograph of 'The Oolacunta As He Appears At Speed'. Because the oolacunta was prodigiously fast and had great stamina, the longest chase extended across twelve miles. By pursuing the oolacunta until 'quite literally, they stopped to die', in temperatures reaching 113 degrees Fahrenheit, Finlayson and his helpers caught and killed nine of the seventeen they sighted.

Other Australian animals thought extinct had been rediscovered, but none after such a long time or involving such a hunt and with so little otherwise known about the creature. Finlayson, with good reason, made much of his success. He had a short account published in the prestigious British scientific journal *Nature*. He wrote a much longer technical piece for the journal of the Royal Society of South Australia, markedly improving on Gould's physical description of the oolacunta and revealing much about its habits and habitat. He contributed a feature to Adelaide's *Advertiser* and Melbourne's *Australasian*, which became the stuff of many news stories and the occasional editorial. When he reprised this account in *The Red Centre* in 1935, it was vital to the success of the book, which became a bestseller.

Yet Finlayson profoundly misjudged the oolacunta's status, failing to appreciate its jeopardy. The last definite sighting was by Finlayson himself on the fourth of his inland expeditions at Ooroowilanie east of Lake Eyre in 1935. In typical fashion, he killed this oolacunta, which is now in the collection of the Museum of Central Australia in Alice Springs. While there have been unconfirmed sightings since, it has generally been accepted that the oolacunta is the only rediscovered Australian mammal species to have become extinct. Although Finlayson prided himself on having achieved a 'startling example' of the 'resurrection' of a species, his hunt for the oolacunta between 1931 and 1935 may, at least in a small way, have contributed to its demise.

Finlayson began these expeditions in the best of conditions. While 1931 has not been classified as a La Niña year, the Diamantina experienced its biggest floods since 1918 and, Finlayson reported, vegetation returned on a 'supernormal' scale after a seven-year drought, with all mammals undergoing a 'quick increase' and rodents in 'plague proportions'. Long-haired rats were readily found near Lou Reese's homestead at Appamunna waterhole on Minnie Downs, where Jimmy Naylon dug out specimens during the day and explained to Finlayson the difference between their tracks and those of other small mammals, as well as much else. There were many more long-haired rats in the sandhills on Clifton Hills where they had first been recorded in 1885. While a few could be seen during the day, they were abundant at night. As Finlayson skinned the oolacuntas, one-handed, he could not stop the rats 'tackling the skeletons and skulls—in spite of arsenic'. They even made off with one oolacunta skull. 'Never recovered it, blast 'em', Finlayson fumed.

Finlayson was able to put his observations in context through Jimmy Naylon and Lou Reese, who kept in touch with pastoralists to the north and south. Their reports allowed him to chronicle, with unprecedented precision, the great fluctuations in the numbers of rats within an extended irruption. Finlayson learned the rats were 'very sparse' as usual on Minnie Downs at the start of January 1931 when they began increasing slowly. From April, their numbers rose sharply, reaching 'plague proportions' in May 1932. They then plummeted, returning to their 'normal sparseness' by September 1932, and remained so until July 1934 when they again rose sharply, but only so they became 'plentiful', about half the numbers of what Finlayson considered a 'plague', and then plummeted again from September.

Winton in Queensland was, again, a centre of these irruptions. Its rats may have come from further west, where they were 'very bad' in 1930, though they were also said to have travelled up the Diamantina from the south. Either way, a few reached Winton in February 1931. By late March, they had arrived 'with a vengeance' and remained for seven or eight months. The record reported kill for one night was thirty-eight—small compared to Ching Foo's wheel-barrow of 142 from the Corfield and Fitzmaurice Hotel fifty years before. When there were no dingo attacks on sheep that spring, one explanation was that the rats were providing the dogs with 'very easy meat'.

For the first time, fish were identified as predators of the rats. One report involved waterholes around Winton. When the rats attempted to swim across these holes, they were swallowed by large yellowbelly, known otherwise as golden perch, which died from being 'unable to digest the unusual fare'. Another report came from around Cloncurry where a fishing party found an adult rat in the gut of an old-man catfish. These reports are consistent with more recent science confirming that an array of fish eat small mammals when they attempt to cross water. Catfish have been identified as 'opportunistic omnivores that will ingest a wide range of different food types' and, at least episodically in the Pilbara, a significant proportion eat the native spinifex hopping-mouse.

The long-haired rats arrived in Longreach where the local health inspector warned, with no apparent basis, that, if unchecked, their damage was 'likely to average £1 per minute'. The rats were also on the downs around Hughenden for months, on stations around Boulia on the Burke River, on the Bulloo where they ate dead sheep and fish on the riverbanks, and on the Barcoo, where they were identified immediately as *Rattus villosissimus*. They appeared too around Julia Creek, where one landowner regretted poisoning them because that meant he also killed kites, which were his station's 'most useful scavengers'. As cat plagues again followed, at least one pastoralist outside Longreach turned against them. In one day, William Crombie of Maranthona Station and his men shot seventy-three cats around the homestead.

The rats' biggest irruption was in the Northern Territory, around Rankin River on the Barkly Tableland, where they remained two years, engendering a new crop of stories. One was of a charcoal cooler—an early form of refrigeration—always carefully closed, especially at night, 'but every morning there was a rat inside'. After several weeks, 'it was found the rats, which first had to climb on top of the cooler, burrowed down through the charcoal and underneath the floor of the inside chamber, then gnawed their way through the wood and then inside'. Another story involved a billiards table. When the balls were left out, 'the menfolk of the house accused each other of getting up in the middle of the night to practise shots because they could hear the click of the balls', then 'discovered the culprits were rats'.

Long-distance travellers reported more. One of the 'strangest experiences' of two motorists circumnavigating the continent late in 1933 'was an all-night drive through a plague of rats near Rankin'. Six months later, two nurses from the Presbyterian Church's Australian Inland Mission encountered this plague from Newcastle Waters to Lawn Creek, a distance of about 380 miles. The rats were particularly numerous around artesian bores—a mark of how the settlers' transformation of the land was altering where the rats were to be found. When Father Long, a Catholic priest based in Alice Springs, said Mass at the Brunette Downs cattle station in the Barkly Tableland while doing the rounds of his 300,000 square-mile parish, 'huge rats were there by the thousand—in the open and in the houses'. According to another traveller, 'it was the theory of the natives that the swarms continued north...and eventually perished through the influence of the salt water in the delta country of the Gulf'.

A party of scientists from the Australian Museum, on an expedition to collect trilobites, also saw hundreds of rats as they drove at night between Mount Isa and Camooweal in Queensland. Ellis Troughton, who had succeeded Edgar Waite as the museum's curator of mammals, wrote of 'advancing hordes...denuding great areas of grass, even destroying the roots, and threatening the survival of valued shade trees by gnawing away the bark, like rabbits in the drought'. The traps set by Troughton around artesian bores 'yielded splendid catches'. When Troughton collected a 'fine series' just within the Northern Territory at Avon Downs, possibly the station worst affected, he was struck by the response of Aboriginal women. They were 'much amused' at how Troughton pickled some of his catches, a process they considered 'quite mad as the rats would be no good for eating later on'.

The account that reached the greatest audience came from the writer Ion Idriess after he spent the first months of 1935 travelling around the interior collecting material for *The Cattle King*—his book about the pastoralist Sidney Kidman, who may have encountered the irruption of 1886 near Cobham Lake. In an interview with the *Sydney Morning Herald* about plagues that blighted the interior, Idriess spoke in terms partly reminiscent of Louis de Rougemont as he described how 'countless millions of rats' had 'gone over the sandhills spread out like a carpet eating everything before them', stripping the bark of young mulga trees, ringbarking them, and eating out the roots of saltbush and cotton bush. 'When they are there, the ground seems to be moving,' he declared. 'If you happen to be in a car they won't get out of the way, and the wheels pass over a crunching mass of rat.'

Idriess was even more concerned about the cats that followed the rats, alive to their destructiveness. He described seeing 'countless thousands upon thousands of these cats', which had 'grown much larger than the ordinary domestic cat'. He declared their proliferation a 'tragedy' because they were 'thinning out the bird life' on the Diamantina and Georgina rivers and Cooper Creek, so that in many places birds had been 'practically wiped out'. While he claimed that a station owner on the Diamantina had shot 130 cats in one night, he considered that 'a small number' given their abundance.

Idriess is now a largely forgotten figure, but Brisbane's *Sunday Mail* identified him in 1935 as 'the most widely-read of contemporary Australian authors due to his extraordinary knowledge of life in wild Australia'. His celebrity saw his interview with the *Sydney Morning Herald* republished by dozens of other newspapers as a confronting piece of realism. The weekly *Land* observed: 'A carpet of live rats, covering miles of country, sounds like a fearsome nightmare; but Mr Ion Idriess...says he has recently seen such a carpet in the sandhills of the Lake Eyre Basin, and he is a witness whose statements may be accepted without hesitation.'

The prime dissentient was Randolph Bedford—a journalist, editor, broadcaster, mining speculator and member of the Queensland parliament representing the state's far south-west. While Bedford recognised that overstocking was destroying the land, he believed that damming rivers such as the Cooper and the Bulloo could facilitate better exploitation of it. In a letter to Brisbane's *Daily Standard*, Bedford denounced Idriess's 'wild statements'. Bedford declared: 'London magazine editors have told me that they did not want me to write of the real Australia but of the Australia as their readers imagine it. Of course, I refused. Apparently Australian city dwellers are to be given similarly false pictures of the continent they don't know.' The *Kalgoorlie Miner* agreed, titling the interview: 'Tall Story from Ion Idriess'.

Idriess soon inflated his experiences further. In a newspaper series called 'Strange Things I Have Seen', there were even more shades of de Rougemont in Idriess's account of how 'five ravenous columns' of rats, 'miles apart, stripped a corner of south-west Queensland'. 'Nothing could stop them', Idriess maintained. 'Horses and cattle galloped from their path; kangaroos and even dingoes kept clear.' But Idriess also suggested that dingoes 'would eat their fill from the flanks of the hurrying millions, snapping a rat, then leaping clear as a dog snaps a piece of meat thrown on an ant bed', and birds of prey and snakes also 'levied their tribute'.

Finlayson did not mention Idriess when, almost a decade after first encountering the long-haired rat on his expedition in search of the oolacunta, he published a five-part series of articles in the journal of the Royal Society of South Australia about the mammals of the Lake Eyre Basin. But in the course of these articles, Finlayson implicitly rebutted Idriess and several other newspaper stories from the 1920s and 1930s, when accounts of old irruptions were particularly hyperbolic including one of settlers in Queensland killing 40,000 rats in a night. Finlayson also explicitly responded to some of the readily accessible accounts of the rat in scientific publications starting with John Gould. Finlayson's thirteen pages about the rat, split between two of his articles, provided the most reflective, substantial account of it.

Finlayson implicitly questioned whether the rat's long guard hairs were an appropriate basis for identifying it, since they were 'not much more marked' than those of some other Australian species. He confirmed that, while 'generally herbivorous', the rat was sometimes carnivorous. He had dissected one that had 'definite traces of animal matter' in its stomach, while Lou Reese and Jimmy had told him that the rat was 'an active and effective predator' upon smaller mammals, pursuing the native plains rat and the introduced mouse, and possibly ground birds.

Finlayson also provided the first description of the rat's burrows, which were 'almost always situated on the slope near the base of a sandhill'. Having opened at least one, he reported that 'the main drive extended 10 feet or more, obliquely towards the centre of the mound, reaching at the end a depth of 3 to 4 feet, below the sloping sand surface'. He recognised that, as with many other species in arid environments, such burrows were vital to the rats' survival, providing protection from predators and from climatic extremes. The centre of the burrows 'was deliciously cool on days when the shade temperature reached 118 degrees Fahrenheit after several weeks of very hot weather'.

Some other species of rat tend to be predominantly male when they multiply rapidly, and males outnumbered the females by more than two to one in Finlayson's sample. In *The Wild Animals of Australasia*, Albert Sherbourne Le Souef and Harry Burrell reported that, during plagues on the Diamantina, all the dead rats seen by a local landholder and all the rats killed by Sidney Jackson and Harry Cottrill were male. Jackson had explained the rats were 'eating each other, and…the females went first', prompting Le Souef and Burrell to conclude that females predominated while there was an 'abundance of food and easy conditions', whereas males predominated in 'conditions of starvation and stress'.

Finlayson thought these figures incredible, and he was right in relation to the Diamantina in 1918. Jackson's diary reveals that most—but not all—of the rats that Cottrill and he killed were male. Little realising that Jackson had engaged in hyperbole, Finlayson looked for a different explanation. Disregarding Jackson's scientific credentials, Finlayson attributed the reports published by Le Souef and Troughton to mere bushmen. When he tested some bushmen around Lake Eyre, he found 'comparatively few' could 'sex rats reliably'. He implicitly ridiculed how they mistook the clitoris of females for the penis of young males, causing them to identify female rats as males. A later scientist James Carstairs recognised that 'female rats are easily mistaken for males by the casual observer'.

Finlayson's discussion of the role of *Rattus villosissimus* in earlier irruptions was particularly significant given the conclusions of Frederic Wood Jones in the 1920s in *The Mammals of South Australia*. Wood Jones had identified the 'hordes' involved in most rat plagues in the Australian interior as *Rattus rattus*, ignoring or overlooking Sidney Jackson's work on the Diamantina. Finlayson persuasively declared *Rattus villosissimus* as 'of considerable general biological interest as constituting, par excellence, the migratory horde rat of arid Australia'. He had no doubt 'the majority of the earlier references to swarms of *Rattus norvegicus* and *Rattus rattus*...and of other large unidentified murids within the above area' involved *Rattus villosissimus*.

Finlayson was otherwise sceptical about these old reports. He argued that terms such as 'horde' and 'swarm' should be viewed cautiously because they 'conjured up a picture of dense masses of rodents moving purposively over the land in plain view', when the rat was nocturnal. Finlayson also questioned whether there had been mass migrations of the rat at night, doubting the many accounts of 'rapid movements of large bodies of animals along... narrowly restricted routes', suggesting they only followed river channels 'on a minor scale'. While he relied on his own network of settler informants led by Lou Reese, he assumed that imagination coloured the accounts of most bushmen.

Since the 'obvious' reason for the rats to move was to leave 'an overstocked and eaten out originating area in search of new and better feedings grounds', Finlayson could see no reason for rats from Queensland to go down Cooper Creek to Lake Eyre when it would take them 'towards country...usually in poorer condition than that which had been left'. His hypothesis was that their spread was 'a slow diffusive drift' due to many small populations radiating out in all directions. He suggested these populations could either be the result of a local increase triggered by abundant food or they could be of distant origin. Although Finlayson questioned the use of the term 'migration', he adopted it in this context because the rats came from elsewhere. He wrote of a 'migration wave' that often appeared suddenly.

A key question was the relative size of the irruptions observed by Finlayson and his informants in the far north of South Australia in the first half of the 1930s. Finlayson wrote that the irruptions of the early 1930s were 'of a phenomenal kind and the aggregate numbers involved were undoubtedly enormous'. But the scant contemporary reportage suggests that these irruptions were smaller than some of those of the past, especially the plague of 1886. Finlayson implicitly acknowledged this when he observed that, though the rats arriving at Lou Reese's homestead on Minnie

Downs in July 1934 were travelling two miles a day, he doubted whether any of the local increases in the first half of the 1930s 'were sufficiently energetic to initiate a wave'.

Finlayson was keenly aware that European settlement was profoundly affecting many species. In *The Red Centre*, he wrote memorably about the impacts of 'the stockman with his cattle, horse, donkey and camel, his sheep and goats and dogs, and the great hosts of the uninvited also—the rabbits, the foxes, and the feral cats'. He observed that 'the results of all this are hailed by the economist and statistician as progress', but had been 'dearly bought', destroying the environment 'which moulded the most remarkable fauna in the world'. But he did not consider how new competitors and predators and the loss of native vegetation were affecting the long-haired rat.

CHAPTER 19

Blitzkrieg

Nature columns were commonplace in Australian newspapers in the 1930s. James Devaney, a Brisbane poet, had two—one weekly, the other monthly, both syndicated—which he relied on for income. Having occasionally written about the long-haired rat, he returned to it in mid-1939 after his counterpart in the Melbourne *Argus* published a series of reminiscences of rat plagues. In a column in Mackay's *Daily Mercury* and Townsville's *Daily Bulletin*, Devaney republished extracts from these reminiscences and asked: 'Do enormous swarms of native rats still occur in Queensland?'

His answer came within a few months. In the wake of heavy rains due to successive La Niñas, the rats irrupted in Queensland and soon spread into South Australia. Already in 1940, this irruption was declared 'the longest on record' in Queensland. In 1941, it was said to have 'never been equalled in numbers or length of stay'. As the plague extended into 1942 because of exceptional enduring rain, that became almost true, though the extent of the terrain occupied by the rat was relatively modest. With good reason, Queenslanders would remember what occurred as a 'super plague'.

The result was unprecedented fear and trembling—probably fueled by the world being again at war, intensifying the horror at this 'threat' and 'invasion' on the home front. Yet this irruption was also the subject of unusually careful observation and precise record-keeping by pastoralists. While no scientists studied the plague in the field, specimens of the rat were sent to the city for identification and analysis. After decades of the rat being of only local concern and interest, Alistair Crombie, a Queenslander at the University of Cambridge, brought it to international attention through the British scientific journal *Nature*.

The largest were up to fourteen inches long and three inches high, 'fat and pudgy', 'as big as condensed milk tins', when they became news in March 1940. By some accounts, they were travelling ten to twelve miles a day into South Australia. By others, they had advanced 110 miles in three months, or about a mile a day, leaving them still 200 miles from Marree. Ken Crombie, a mailman on the Birdsville Track, best known for his appearance in Ernestine Hill's *Great Australian Loneliness* of 1937 because she travelled in his 'two-ton truck with its engine-fittings held together by safety pins and bits of string', was widely quoted about the rats. He reported that, while primarily nocturnal, the rats 'sometimes were seen during the day in large batches', followed by many dingoes. Because rats were 'digging up a lot of ground in search of roots and other growth for food', Crombie 'could not see any trace of the road in places'.

The rats' arrival—and the damage they might do—was dreaded

around Hawker in South Australia's northern wheat belt. But to reach there, the rats had to cross Cooper Creek when it was 'a swirling, fierce stream wider than at any time since the country was opened up' and 'negotiate practically 500 miles of practically uninhabited country'. They did not do so. After many drowned in the Cooper, the survivors headed west for Lake Eyre as, contrary to what H.H. Finlayson proposed in 1939, they often followed watercourses.

In western Queensland, there was initially little concern. A station hand observed: 'People in these districts are afflicted with so many worries that they only regard the rats as a minor affair.' The chairman of a regional pastoral board declared the 'natural tendency' was for the 'plague to die out of its own accord'. But then the irruption grew and grew—especially around Boulia, which had experienced earlier plagues but had not been their centre until August 1940. A report of two 'distinct' types—one black, the other white—ignored how some long-haired rats are albino. Another report recognised all were long-haired rats. When a specimen was sent to Sydney for examination in case it was *Rattus rattus* carrying bubonic plague—a possibility taken seriously because there had been cases in Charters Towers in 1900 and Aramac in 1920–21—it was identified as *Rattus villosissimus*, implicitly clearing it of any connection with disease.

Heber Longman of the Queensland Museum, who had written extensively about Australia's rats, offered to examine specimens, eager to be involved. When the Boulia correspondent of Brisbane's *Courier-Mail* sent five, Longman advised they too were long-haired rats. 'Today's parcel is of great scientific value,' he declared in accepting them into the museum's collection in August 1940. 'In previous plagues we have vainly sought for specimens.' Since all five were male, he requested some female. After receiving two—one dark, the other again albino—the museum displayed them that October.

Some, thought to be travelling from the south-east, reached as far north as Hughenden on the Flinders River but 'not in any way in sufficient numbers to constitute a plague'. They also increased around Winton where a pastoralist identified as many females as males and, to the puzzlement of local residents, 'it was common to see a property practically eaten out while a property adjoining was left almost untouched'. Their numbers were greatest in and around Boulia where ninety-five per cent were reported to be male and the stench from dead rats under buildings was 'almost unbearable'. By one account, the rats 'ravaged Mitchell grasslands over an area of more than 20,000 square miles in Boulia and Birdsville districts'.

The pastoralists' prime concern was to protect their homesteads. J.G. Scholefield of Paton Downs—a sheep station outside Boulia—attracted most attention for the way he 'killed and counted'. His tallies around his homestead were 1076 rats on the night of 6 September; 1434 on 7 September; 742 on 8 September by turning off the water supply and leaving out two tins of water containing arsenic; and 460 on 9 September. By the start of October, Scholefield had killed more than 10,000—a figure later inflated to 100,000. Each morning, he had the corpses placed on a lorry—a sight photographed blurrily for Townsville's *Daily Bulletin* after a record nightly kill of 1648. Then one of his men dumped the rats several miles away.

Most of the rats' behaviour was familiar. Their tracks 'effaced all hoof marks in a cattle yard'. Some trees were 'eaten bare of bark'. Storekeepers 'suffered heavy losses'. Other assaults were new. Having previously been said to eat their way through doors, the rats 'gnawed through homestead walls' and 'prised lids off seven-pound tea tins'. They burrowed five to six feet under a shearing shed and got through the timber. They ate 'wool off the backs of sheep'. They ate solder, 'causing water tanks to leak'. A live calf had 'one ear partly eaten off, and the skin chewed off two legs'. They ate the trousers of Longreach's Catholic priest.

The most sensational reports came from the *Courier-Mail*'s Boulia correspondent. He quoted a local selector who, like Louis de Rougemont, claimed to have watched 'a mob of rats, estimated at 200,000, literally march across his property. When they had passed, hardly a blade of grass was left. The property was like a ploughed field.' The *Mail*'s correspondent maintained that the rats 'sickened' those plants that survived. He also invoked the threat of bubonic plague—recording how one of his friends hit a rat with his hat, then found his hat covered with fleas. His stories of cats underlined the danger: 'When I first came to the district, Boulia was alive with rats, now there is scarcely a cat alive. Residents here believe that the rats carry a disease that is deadly to cats. In one instance which I can vouch for, there were eight cats attached to one home. Now there is none.'

Many reports cast the rats as a natural disaster. 'In the Boulia district, where a short time ago, the seasonal outlook was good, the rat plague has caused a complete change', began one. 'Unless it vanishes shortly all feed will be ruined. With the fouling of pastures by this plague, stock have gone off greatly in condition.' A local stock inspector maintained that the value of the country had fallen by 'a tremendous extent'. But Queensland's chief stock inspector disagreed. He suggested that, 'as the plagues came only once in ten years and lasted only a few months', they should be 'endured', especially since the rats always followed 'an unusually good season', which compensated 'for any damage done by the rats'. Others reckoned it useless to kill the rats since they were 'as numerous as ants'. As usual, Aboriginal people considered the rats a boon. When the flying doctor visited Birdsville, a group of Aborigines showed him a pot full of rats, stewed for dinner.

Elsewhere, there was scepticism. A West Australian newspaper quipped that a report of the rats being followed by hordes of snakes and goannas sounded 'like a "rum" fantasy, with the pink elephants missing'. The Melbourne *Argus* featured a 'seasoned bushman' who 'did not believe in Queensland's rats'. Having experienced several plagues of the introduced mouse, this bushman suggested Queensland 'was suffering a severe mouse plague'. Eventually, the *Argus* implicitly acknowledged its mistake with a story titled 'Rats Are Rats', an account from a 'harassed resident' of Rankin River about the horrors of combatting the 'hungry invading hordes'.

When the regional air ambulance visited Boulia for the first time, in October 1940, a man and dog were put on guard through the night to protect the aircraft's fabric. Yet there were signs the plague was diminishing. 'The rodents have practically disappeared', ran one account that November, and so they largely did around Boulia. J.G. Scholefield linked the rats' disappearance from Paton Downs to unusually heavy cyclonic rains in January 1941—the first time that big rains were identified as a cause of the end of a plague. Scholefield reported that these rains caused very high 'infantile mortality' among the rats, with the 'newly born unable to survive the trying weather conditions'. In their wake, another pastoralist found dead rats stacked to a depth of twenty-seven feet in a well.

The rats also appeared in Winton, Cloncurry and Mount Isa as the exceptional rains gave rise to one of western Queensland's longest floods, from January until June 1941. While the Brisbane council had upped its rat bounty to a shilling during an outbreak of bubonic plague in 1921, a twopenny bounty offered by the Cloncurry council elicited a big response, with a hunt at the local railway station bagging nearly 200 in an hour. In Mount Isa, where the council also paid twopence, its rat man cried, 'Bring along your dead!' then incinerated the corpses of these 'big grey voracious-looking rodents'. Although some men joined the 'rat crusade', with the largest daily cheque about £5, those who responded were primarily boys quick to convert their earnings into meat pies and humbugs.

Longreach, on the Thompson River, was most affected. From early January 1941, townspeople found dead rats in the streets each morning. While some were killed by cats and dogs, the local health inspector, L. G. Homer, identified most as dying of natural causes. But as the plague continued, men and boys and sometimes women formed 'rat gangs'. Even with no incentive from the local council, they went on the attack each night with 'sticks of all sizes and shapes of weapons'. A hotel reported a catch of 236, far eclipsing that of Ching Foo in Winton in 1880. Local residents were also keen to spread poison, but inspector Homer advised against doing so, as the rats would probably die under buildings.

The townspeople feared that when the flood receded, thousands if not millions of rats on the far side of the Thompson would invade Longreach. When inspector Homer travelled by train to investigate the threat, he stood at the front of the engine and, with the fireman's poker, knocked hundreds of rats from the line into the water. He reported that some had burrowed under the track's wooden sleepers and, in one burrow, there were twelve young rats. A local resident wanted a moat dug around the town and filled 'with poison to stop the progress of the rats'. Inspector Homer advised against it, this time because 'large number of native birds would also suffer'. Instead, he arranged for the council to borrow a flame thrower to destroy those which townspeople would kill when the rats arrived. It proved an unnecessary preparation as no invasion occurred when the water receded that February. But the rats remained common on nearby stations where stockmen stopped killing snakes because of their importance as a rat control, just as motorists, who usually delighted in running over goannas, began to 'go out of their way to avoid them'.

The response of children to previous plagues had gone unrecorded. During this irruption, the *Longreach Leader* published a 'Children's Corner' of up to three pages every Saturday. As the rat loomed large in their lives, children wrote about what it attacked, how they killed it and what they did with the dead. A boy on one station reported: 'The rats are thick. We shoot them with a little pea rifle and caps. Gee, they squeal when they are hit.' A girl on another station wrote: 'I have found a new way to catch a rat. I tie a bone on a piece of string, hold the other end in my left hand, I sit on the steps behind one of the house blocks and when the rat comes along I dong him. It sounds cowardly but I must get pennies somehow.' A boy wrote from Barcaldine: 'I killed one. I wouldn't have killed it only I had 119 pieces of corn up and it ate the lot. So the sentence was death. Last Thursday we were playing hockey with a rat for a ball.'

Much of the focus was on the far north. In early 1941, the rats extended from eastern Queensland to west of the Barkly Tableland in the Northern Territory, where a drover claimed the rats were biggest and 'only males were observed, the females staying in their burrows…to be fed by the males'. For weeks, Charters Towers in

Queensland felt it was being 'menaced'. But most of the rats were stopped by the swollen Burdekin River, which soon swirled with carcasses, and the local health inspector, J. G. Roberts, dismissed reports of significant numbers of the rats reaching the town as 'hallucinations'. His proof: there had been no increase in complaints, requests for rat baits or sales by chemists of rat poison.

There were also tentative reports that the rats might have extended to Townsville—the only accounts of them possibly reaching Australia's east coast. Rodents, said to fit the description of the long-haired rat, were reported to be 'more than usually prevalent' on Townsville's outskirts in early March. A few days later, carcasses washed up on Townsville's foreshore were again said to bear 'certain resemblances to the western pest'. Those examined by the health inspector were so decomposed that identification was impossible, but 'well-informed' residents thought that 'remnants of the plague which played such havoc in the western areas' had reached Townsville.

World War II shaped language. Australians embraced rats as something to be lauded—celebrating their soldiers in Libya as 'Rats of Tobruk' after the Nazis branded them 'poor desert rats'. A story about the long-haired rats possibly again heading from Queensland into South Australia described them as seeking 'Lebensraum', the prime Nazi term for territorial expansion. When this story first appeared, it was titled 'Rat Krieg'. Another report about the rodents approaching Charters Towers was titled 'Rat Blitzkrieg'.

Market gardening was in decline due to the cost and shortage of labour. One reluctant gardener was Sang Chong, a Chinese-Malay man who responded to recurrent prejudice by identifying as British having been born in Singapore. After trying vainly to sell his garden outside Cloncurry in the late 1930s, he was benefitting from the big rains—producing tomatoes weighing up to two-and-a-half pounds—when the rats attacked. His response was to buy cats, for sixpence per head. 'Watch for poor old Chong on his rounds of the town on Wednesdays and Saturdays,' he advertised, 'and get rid of your surplus cats at a profit.'

Outside Longreach, the country 'was alive with foxes', with their pads even more common than rat pads in some areas, alarming many pastoralists. They recognised the foxes were feeding on the rats, and anticipated correctly that the foxes would switch to sheep when the rat plague ended, prompting one pastoralist to kill 130 foxes and another to kill fifty.

A few pastoralists thought they benefitted from the plague. They credited the rats with reducing instances of flystrike, inhibiting growth of the Bathurst burr, which otherwise got into wool, and providing easy prey for foxes and wedge-tailed eagles, both of which otherwise ate lambs. Most pastoralists saw only costs. The rats created 'almost continuous' work by undermining fence posts. If they did not kill newborn lambs, as first claimed around Winton in 1880, they disturbed pregnant ewes, causing some to abort their young. The rats' deaths within homesteads necessitated 'a constant use of disinfectant, pulling up of floor boards, and crawling about under houses'.

The rats' alleged responsibility for bushfires attracted most attention, prompting several pastoralists to conduct their own experiments and publish their results in the rural press. One pastoralist could not induce the rats to eat wax matches even when he covered them with food. But when two others put wax matches in dry grass, they caught fire. E.C.C. Luck of the Portland Downs merino stud was most ambitious—experimenting for six weeks with rats in petrol cans, drums and cages. He concluded that the risk of them igniting matches 'was so small it was not worth considering'. Instead, he contended that ammonia in the rats' urine and droppings caused the grass in their nests to combust spontaneously and become 'miniature volcanoes'. This explanation was widely embraced even though an agricultural chemist dismissed it, prompting *Queensland Country Life* to announce, for once, 'The rat is innocent.'

When cats again followed the rats and died in large numbers, their deaths caused more puzzlement. 'Plain starvation', 'canine distemper' or 'the hairs of the rats forming balls in their intestines' were among the suggested causes. That the rats themselves might be sickening was both welcomed as an end to the plague and feared as a source of infection. When it was said the rats had developed a disease that 'strongly suggested leprosy' because it 'caused loss of fur, outbreak of sores on their bodies, and rotting off of parts of their tails', scientists explained the rats were probably lacking in 'some vital vitamin'. There was no need for panic since 'leprosy in rats had no bearing on human disease'.

The rats also irrupted in the far north-west of New South Wales, ignored by the local press but recorded in October 1941 by the manager of Tibooburra Station in a letter passed to David Fleay, the director of the Healesville wildlife sanctuary outside Melbourne who had become the nature columnist for its *Argus*. The manager described how the rats had been present 'in plague proportions' for two years, resulting in 'great destruction of feed' while attracting many dingoes. Oblivious to their identification in western Queensland as *Rattus villosissimus*, Fleay suggested they were *Rattus rattus*.

Large numbers remained in Queensland through much of 1942.

A 'semi-plague' continued in and around Longreach where the shire council tried to control it by providing baits free of charge. The rats were thought 'to blame for much of the sickness' in nearby Ilfracombe. They reappeared around Muttaburra, re-entered Winton and were a nuisance around Hughenden, though at least one gardener realised that elevated vegetable beds in old iron tanks and forty-four-gallon drums were a solution, enabling her to harvest cabbages, silverbeet, carrots, beetroot and lettuce.

The rats were still prevalent that September in western Queensland, especially around Julia Creek and Richmond, but it was 'generally agreed' they were disappearing because the land was increasingly dry, while feral cats and foxes were also dying from what one government veterinarian identified as canine distemper. As gardens stopped being pillaged that October, a girl wrote to the *Longreach Leader*: 'Dad planted some rockmelons and watermelons and they are all up. I love both. They should be ripe for Christmas. We get eight or nine eggs a day. I hope the rats don't come back.'

Dingoes excited particular concern, as they feasted on the rats and their numbers escalated, despite the settlers' usual attempts to bait them with pellets of strychnine inserted inside small pieces of beef. One theory was that the dingoes 'could not be poisoned while they had food galore around them', because they ignored the baits. Since the rats were thought capable of almost anything, it was also speculated that they got to the baits first and ate the beef, while rejecting the poisoned pellets that the dingoes then ignored because the rats had been at them. Whatever the cause, the dingoes became 'menacingly numerous'.

The one scientist to analyse the rat plague was in England. Alistair Crombie, a zoologist particularly interested in population dynamics, who would become much better known as an historian of science, studied the rat for reasons personal and professional. He almost certainly experienced the irruption of 1931 on Maranthona, his family's sheep property forty miles north of Longreach, while on holidays from boarding school. Then, while he was at the University of Cambridge, Maranthona was infested again from 1940 until 1942 and Crombie's father William grappled with the plague as a member of the Longreach Council and pastoralist.

'Rat Plagues in Western Queensland', published at the end of 1944, was Crombie's third piece in *Nature*. His data from Maranthona, compiled by his family, was exceptionally strong; his general understanding of the irruption was weak. While the conventions of academic science still allowed Crombie to describe the rats as 'vermin'—a subjectivity now impermissible in most scientific journals—he felt obliged to hide his connections with western Queensland. Instead of crediting his family with chronicling the plague for him, he acknowledged the assistance of 'correspondents…nameless, but not unthanked'. Rather than identify Maranthona, he wrote about a point which he called '*X*'—perplexing readers as to why there should be such a wealth of rat recording at and around this spot and why Crombie should have access to it.

The most striking feature of Crombie's account was, again, the fluctuations in the plague. He reported that in October 1940, they were sixty miles north-west of *X* or Maranthona. That December they reached *X* itself 'in fair numbers'. In January 1941, 'their numbers greatly increased and persisted at this level until about October of the same year', with between eight and fifty-three caught at night in traps set around the verandahs of the homestead. 'The population decreased suddenly in December 1941, rose again in February 1942, and fell once more in April.' Afterwards, 'rats continued to exist at their normal low level of population', indicating they were always there.

Not surprisingly, when Crombie explored causes of the irruptions, he found they generally occurred in years that were exceptionally wet. But he also observed that 'other wet years did not have plagues', implicitly recognising there was much left to explain. The end of the plague was one example. Unaware that J. G. Scholefield had linked the disappearance of the rats on Paton Downs to the exceptional rains of January 1941, Crombie noted a similar correlation for Maranthona sixteen months later. With typical specificity, he reported, 'The rats finally disappeared after the heavy soaking rains of May 26, 1942.'

CHAPTER 20

Taipan

Queenslanders delighted in the end of the plague. 'No rats in the bush this year', a newspaper announced in 1943, but their disappearance was short-lived. A small irruption extended from 1944 into 1945. A much bigger one started in 1949 when the Channel Country was inundated and the floodwaters of the Cooper stretched fifty miles wide and approached Lake Killalpaninna. This plague continued in 1950 when there was even more rain. Lake Eyre filled, perhaps more than it had since the arrival of Europeans, and the rats reached it. When a team of armed servicemen was asked in 1951 for their 'most vivid impression', after completing a 9000-mile, five-month, 150-town lecture tour of Queensland, they answered 'rats'.

This time the biggest were said to be as large as small kittens, up to four inches tall, and weighing three-quarters of a pound.

When fish died in waterholes around Boulia after a heatwave, the rats reportedly swam out to eat the fish, rather than being consumed by them as on previous occasions. Lorry drivers could not stop for half an hour without rats boarding their vehicles and 'ripping the feed bags to pieces'. As usual, rats were accused of causing fires by igniting wax matches dropped by settlers, though there was some acknowledgment that bushmen who failed to put out their camp fires were the 'greatest menace'. The rats' impact was felt particularly when many areas returned to drought. 'With the plague of rats contaminating what herbage there was...the picture was sordid and tragic for stockowners,' the *Longreach Leader* reported.

Over the next twenty-five years, the rat was studied by a new generation of scientists, who sometimes became museum curators but typically worked elsewhere. Many joined the Commonwealth Scientific and Industrial Research Organisation (CSIRO), which became Australia's pre-eminent zoological and ecological research organisation after establishing a Wildlife Survey Section in 1949. Others worked at universities, which came to receive much more research funding. These men—and occasional women—not only engaged in extensive fieldwork but also, for the first time, examined the long-haired rat in laboratories. Some of their research was a response to the rat pillaging crops. Much of it had no immediate economic imperative. The result was a quantum leap in understanding of the rat including the rediscovery—more like the discovery—of another snake closely linked to the rat. Because of the toxicity of its venom, interest in this taipan was intense.

The CSIRO first studied the rat during a plague in the mid-1950s that included small outbreaks in the Northern Territory near Alice Springs and on the Barkly Tableland but was largely confined to Queensland. Dundee, a sheep station south of Richmond, was particularly affected because it was the site of an agricultural experiment conducted by the University of Queensland in which a small crop of sweet sorghum was being grown to supplement the native Mitchell grass. When the rat was attracted to the sorghum—and devoured it—George Dunnet, a Scottish zoologist working for the Wildlife Survey Section, went to investigate. With the local plague at its peak, he found burrows every four to five paces, or 300 per acre, on part of Dundee, much like Edward Palmer in the Gulf Country in 1870. Because the rats were under intense stress, they sometimes came out during the day, providing diurnal birds with prey. Dunnet saw fork-tailed kites with rats in their talons and discovered a wedge-tailed eagle's nest containing several rats, while the ground below was littered with castings full of the rats' remains.

The next big plague, following a drought lasting eight years in some places, began in the far north in 1966 and intensified after unusually heavy summer monsoonal rains. News reports suggested that the rat moved slowly south along the Georgina River, then crossed the Diamantina in December, and in May 1967 infested the upper Birdsville Track, extending across 'an area 50 miles wide and 150 miles deep'. By these accounts, the rat was still moving south, and breeding as it went from Clifton Hills, travelling an estimated 'three to five miles a day' as it headed towards Marree. As the rat excited considerable alarm, Queensland police warned members of the public against camping out.

The rat was also found in many other places. It was reported for the first time in Western Australia—south of Hall's Creek, across the border from the Northern Territory—and north of the Barkly Tableland at Katherine and along the Humbert River. It was also recorded 135 miles south-west of Alice Springs and in South Australia again on Clifton Hills, where H. H. Finlayson had encountered it while searching for the oolacunta in 1931. The station's manager identified Clifton Hills as another area where the rats were always present in small numbers. But during this plague they were joined by many arrivals from the north. Their density was even greater than Edward Palmer and George Dunnet had recorded with 'a hole every eighteen inches over several acres of soft ground'. Another of their prime sites was Brunette Downs on the Barkly Tableland where they had been reported in the 1930s too.

In the only doctoral thesis devoted to the long-haired rat, James Carstairs of Monash University supplemented his own fieldwork with a questionnaire he sent to 1000 pastoralists in the Northern Territory, western Queensland and northern South Australia, eliciting almost 300 responses. Carstairs concluded, contrary to Finlayson, that the rats often followed river channels—because of topography and the availability of food and water. His focus was Brunette Downs where, again, the rat could always be found, according to the station's manager, and Carstairs provided more evidence of how it could boom and bust, year after year, with rain playing a big role in each part of its cycle. After the number of rats first peaked in November 1967, immense rains at the start of the next wet season led to many dying, 'undoubtedly drowned', a pattern repeated in 1968, and then again on a smaller scale in 1969.

Meanwhile, Alan Newsome and Laurie Corbett of CSIRO discovered that when the rat's numbers plummeted towards the end of the plague—and the zoologists' traps failed to catch any—dingoes were still eating them, a clear mark of how the rats can still be present when scientists find none. But when not even the dingoes could catch any rats, only a few switched to scavenging cattle carcasses, while many others succumbed to canine distemper or died of starvation. Corbett suggested that having fed almost exclusively on the rat for years with their pups reared entirely on them, the dingoes had 'such a specialised search image' that many could not switch to other prey.

The strictures of the academic journal article increasingly constrained how scientists wrote about the rat, but Newsome and Corbett enjoyed more freedom in a book chapter about their rodent work. In conventional fashion, they observed: 'The rats were a great nuisance. Cattle and horses had hooves nibbled, saddlery was eaten, and people camped away from settlements were bitten in their sleep.' Much like Sidney Jackson on the Diamantina in 1918, they described how 'to walk at night with a torch was to see rats every few paces scurrying along worn trails to numerous holes, to see them foraging on open grassland, and to hear them squeaking and fighting in every corner.' Unusually, they concluded: 'There was no other word than fantastic to describe it all.'

Laboratory studies, largely using animals caught by James Carstairs on Brunette Downs, which became the basis of a colony at the CSIRO's Division of Wildlife Research in Canberra, complemented field studies of the rat's capacity for reproduction. While the females have twelve teats, their litters average about seven. Blind, pink and hairless at birth, like all rats, they are ignored by their father. They receive little care from their mother, who weans them after about three weeks, four days after they open their eyes. By then, their mother may be preparing to produce another litter since the female has the highest ovulation rate of any Australian rat, can conceive within hours of giving birth and her gestation period is about three weeks. Before long, the female progeny of her first litter may also be pregnant as they can conceive when forty days old.

Other laboratory studies showed that the rats are in some respects well suited to hot, arid conditions since they possess a low basal metabolic rate and have a high ability to tolerate hypothermia. But in other respects the rats are poorly adapted, as they cannot survive on a diet of dry seeds and need relatively large volumes of water, unlike some other Australian rodents. If the rats cannot find free water, they need to secure it from moist green vegetation. If they do not, they will die in about a fortnight.

Another CSIRO team searched vainly for the rats in the Northern Territory in 1971, but was more successful near Boulia in Queensland where residents reported the rats had been there continuously through thirteen years of drought. After encountering the rats in the Simpson Desert in 1972 and then finding evidence of another plague near Winton in 1973, Kent Williams concluded there were many small irruptions that occurred out of phase with each other in response to local conditions and that they often went more or less unnoticed. The scientists also encountered plagues of cats which, eager for refuge from the heat, did not just seek shelter under trees but often climbed them. With manifest amazement, Williams wrote of Boulia: 'We counted thirteen cats in one large lonely tree.'

Scientists from Adelaide encountered the rats in many other places. A team from the South Australian Museum found what the scientists believed were 'permanent populations' at two sites offering reliable water. One was a product of pastoralism—a large free-flowing artesian bore on the western side of Lake Eyre. The other was at Dalhousie, part of South Australia's most northerly group of natural mound springs, fed by the Great Artesian Basin.

Chris Watts and Heather Aslin of Adelaide's Institute of Medical and Veterinary Science found the rat across much more diverse terrain in South Australia's north-east and Queensland's south-west. After trapping it among reeds and sedges around a bore drain, in a sand ridge close to a flood plain and in 'gibber' country littered with polished rocks and pebbles, Watts and Aslin concluded that the rat could be found in all types of habitat as long as there was green growth to provide it with water.

North American mammalogists, J. Mary Taylor and E. B. Horner, provided one of the most illuminating discussions of the rat as part of a systematic reconsideration of all Australian species of *Rattus*, which saw them examine every museum specimen in North America, Europe and Australia. In their 130-page report published at the start of 1973, they paid unprecedented attention to whether the range of the long-haired rat had shifted over time. They concluded it was a creature in retreat, which had 'contracted from the south and south-east'.

Almost immediately, the rat was irrupting again on a vast scale, but without reaching its old terrain identified by Taylor and Horner. After exceptional rains in 1973, the rat spread down the eastern side of the Simpson Desert in 1974, its burrows honeycombing the flood plains and watercourses, while on Farrar Creek it created pads through the Mitchell grass. As the plague continued into 1975, when floodwaters again reached Lake Eyre, filling its southern arm for the first time on record, the rat ranged from Mount Isa in Queensland to Wyndham on the Gulf of Cambridge in Western Australia, to Uluru in the Northern Territory and Woomera and the plains east of the Flinders Ranges in South Australia.

One of Australia's most prominent naturalists, Vincent Serventy, encountered the rat around Lake Eyre. He recorded that a rat on the lake's northern shore 'plunged in, swimming strongly. About fifty metres out it turned parallel to the shore, then began to flounder.' After Serventy waded out to rescue it—handling it with great care, having been bitten by one before—it went back in the water. 'Half-an-hour later a black kite swooped and lifted the swimmer a metre into the air, only to drop it quickly as the rat bit viciously.' A year later, when Serventy was on the lake's southern shore, he found many dead rats, leading him to conclude that 'at least some of these swimmers had made the long journey'.

The end of this plague was a puzzle to scientists—a striking instance of the rat declining despite continuing good conditions, observed Chris Watts and Heather Aslin in *The Rodents of Australia*, the first such book. While Watts and Aslin recognised that the long-haired rat's terrain had diminished in New South Wales, they otherwise cast it as highly resilient to European settlement. Unaware of the scale of the 1880s plague, they suggested that the rat may have reached its greatest extent in 1975. Having seemingly dismissed or doubted old accounts of the rat as hyperbolic, Watts and Aslin now embraced them. 'Camping on the gibber plains towards the end of a rat plague when the animals are starving is an unforgettable experience,' they wrote. 'All the stories of rats in tents and sleeping bags, gnawing hair and ears become quite credible.'

Meanwhile, the raptors that preyed on the rat were the stuff of more study by ecologists and ornithologists, professional and amateur, who focused on the eastern grass owl and the barn owl. But knowledge of the snakes that preyed on the rats lagged far behind. What occurred when Thomas Wall collected what became the type specimen of the long-haired rat on Cooper Creek in 1847 was emblematic. Wall failed to return with any of the snakes occupying the same terrain—perhaps because he was fearful of being bitten, perhaps because they were too hard to catch in the deep fissures in the ground that offered them 'first-rate accommodation'.

Herb Rabig, who owned the Cuddapan cattle station outside Windorah in western Queensland, triggered the most exciting work, having already written about the letter-winged kite and the long-haired rat for Queensland's ornithological journal *Sunbird*. His article was prompted by the appearance of the kite at Cuddapan at the end of the rat plague that started in 1966. While Rabig had heard of one possible previous visit by the kite, he had never seen one in his more than sixty years in the area. Then, in 1969, a flock occupied a fenced area around a tank—'a little oasis with plenty of water' not only 'lush with edible plants' but also home to 'a very dense colony of long-haired rats which had reach starvation point, after finally destroying their own protective cover by denuding the blue bush, salt bush and mimosa'.

Rabig killed a snake during the next rat plague, while on a neighbouring property called Morney Plains. It was of a type he had seen before but, because he did not know its name and, most likely could not identify it in his natural history books, he decided it warranted further scrutiny. While Australia's museums had been eclipsed in many spheres by the CSIRO and universities, they were still at the forefront of identification, classification and collection of specimens, which members of the public often provided. Rabig put the snake's head and tail in a tin and sent them in August 1974 to the Queensland Museum.

One of its curators, Jeanette Covacevich, found two matches after scrutinising the literature. One was *Diemenia microlepidota*, the small-scaled brown snake, named and described in 1879 at the National Museum of Victoria in Melbourne on the basis of two specimens said to have been found in north-western Victoria at the junction of the Murray and Darling rivers. The other was *Diemenia ferox* named and described at the Australian Museum in Sydney 1882 on the basis of one specimen, soon lost, from 'Fort Bourke', presumably Bourke in western New South Wales. That was it for almost a century—a little like the oolacunta until 1931. But rather than being thought extinct, this snake had been ignored. It was a scientific mystery no one had tried to solve.

Covacevich set out to discover more with herpetologist Charles Tanner, a key supplier of Australia's Commonwealth Serum Laboratories, who captured snakes to milk them of their venom so the laboratories could produce antivenene. In September 1974, a month after receiving Rabig's specimen, Covacevich and Tanner went to his station, Cuddapan, uncertain whether they would discover any more. As they approached, they found one, freshly killed by a car. The discovery of a full specimen, if a squashed one, meant the expedition was a success, and it was just a start. Rabig played a big part—guiding Covacevich and Tanner, much as Lou Reese had guided H.H. Finlayson on Clifton Hills. In a week on Morney Plains, Tanner caught thirteen large healthy specimens, partly because spring is the season when this snake is most likely to be found on the ground rather than beneath it. These snakes allowed Tanner to establish a breeding colony that would be the focus of much more research.

In the north-east of South Australia, the Wangkangurru's name for the snake was *dandarabilla*. Covacevich and three colleagues eventually reclassified it as *Oxyuranus microlepidotus*, Australia's second species of taipan. The first, the coastal taipan, *Oxyuranus scutellatus*, was already renowned as the world's most venomous snake. 'Killer of the Cane Fields', Australia's best-known snake man Eric Worrell dubbed it; Australia's 'most glamorous snake', reckoned the deputy director of the Commonwealth Serum Laboratories, John Trinca. The new inland or western taipan from the Channel Country joined and eclipsed it. 'Two Taipans!' Covacevich titled one of her many articles, expressing the excitement at this 'herpetological sensation'.

Its fame was due to its venom. The standard test on mice established the inland taipan's venom as the most toxic in the world, more than twice as poisonous as the coastal taipan, and twenty times that of the king cobra. As a result, the inland taipan became known as 'the world's deadliest snake'. When Australia's renowned 'crocodile hunter' Steve Irwin was filmed with an inland taipan, he would tell his viewers that it possessed 'enough venom to kill 150,000 rats' and 'enough to kill 100 blokes'. Covacevich was quick to emphasise that it was not the most dangerous snake for humans. That accolade depended not just on toxicity but also the quantity of venom in each bite, the length of the snake's fangs, its temperament and habitat—and the inland taipan was generally shy and placid, found where there were few people, injected relatively little venom and had small fangs.

Its links to the long-haired rat emerged gradually. There had been a rat plague when, as Covacevich and her colleagues discovered, an inland taipan had bitten someone in 1967, in the Channel Country. There was another plague in 1974 when Rabig sent in his specimen and Covacevich and Tanner caught many more with ease. But in 1976, when that plague had ended, Covacevich and Tanner found just two, which were starving and soon died. Over the next four years, despite making annual visits to the Channel Country, scientists from the Queensland Museum found none. It was only when more big rains triggered another rat plague in 1981, and the rats were 'literally carpeting the paddocks and grasslands' at night, that the snakes again became 'a common sight'.

Covacevich and Tanner were still cautious about the association between the inland taipan and the long-haired rat. 'Accurate surveys have not been undertaken to prove that the snake populations rise and fall with those of the rat,' they wrote. 'The snake may just be more conspicuous when food is plentiful.' But while the surveys they envisaged have not been done, Covacevich eventually concluded that the taipan has 'a highly-specialised life history inextricably linked to that of the native rat', and other scientists have agreed. When the Queensland Museum moved to a new building on Brisbane's South Bank in 1986, one of its opening exhibitions and longstanding displays was *Feast and Famine* exploring when and how the taipan preys on the rat.

The taipan pursues the rat during the day when it is usually sleeping in its burrows or deep cracks in the ground. Because the rat's teeth could pierce its skull, the snake strikes with great speed, then releases the rat to avoid being bitten. The taipan's venom usually immobilises the rat very quickly. If necessary, the taipan strikes again. Since it is like all other snakes in being unable to bite or tear its prey into smaller pieces, the taipan swallows the rat whole, usually head first, then slowly digests it.

The location of the inland taipan's original discovery by Europeans in 1879—said to be at the junction of the Darling and the Murray Rivers—has generally been doubted because it is so far from the taipan's current terrain in Queensland and South Australia. The sources of colonial specimens were sometimes confused and mis-described. But like many other Australian species, the range of the inland taipan may have shrunk dramatically.

That the rat reached Pooncarie on the Darling in 1887 and was thought to have come close to the river's junction with the Murray at Wentworth, makes it more likely that, in the nineteenth century, the taipan could be found there too.

CHAPTER 21

Small Tigers

In his first book, published in 1894, Henry Lawson wrote of bush cats 'always dragging in things...unknown in the halls of zoology'. Unconcerned by their deaths and happy to deride Australia's distinctive fauna, Lawson branded the cats' prey 'ugly, loathsome, crawling abortions which have not been classified yet—and perhaps could not be'. Lawson still suggested that scientists should try. He advised the 'Australian zoologist' to 'go out back with a few bush cats', as zoologists and ecologists increasingly do, with feral cats one of the great threats to the continent's environment, able to survive in every terrain, almost devoid of predators, and still not the subject of an effective control.

While the number of cats fluctuates depending on conditions, it probably averages at least four million in the natural environment, with some estimates as high as twenty million. Their prey across the continent includes 400 or more vertebrate species. As cats switch between these animals depending on availability, they have become the 'small tigers' the Melbourne *Herald* feared in 1886. Their impact on the long-haired rat during the peak of its irruptions may be of little significance for the rat since it has always been preyed on heavily when abundant. But as cats feed almost entirely on the rat during these periods, they may affect other species such as the letter-winged kite, which have depended on the rat. Cats are also a threat to the rat if they target it at the end of a plague, leaving fewer to survive until the next irruption, or if cats attack the rats in their refuges.

Scientists have grappled with these issues, even as their study of the rat has diminished. Funding cuts have been one problem. As Commonwealth governments slashed the CSIRO's budget and required it to generate more of its own income through consultancies, it reduced its ecological work, then closed its Division of Wildlife Research in 2005. Federal support of ecological research within universities, through the Australian Research Council, has also declined. State governments have closed their environmental science divisions. But opportunities to study the rat have also diminished as its irruptions have become rarer. After the big plague of the mid-1970s, it was fifteen years until the next in the early 1990s, then twenty years until the next in the early 2010s.

One discovery involved another venomous snake, up to 1.8 metres in length. As is often the case, there was a long gap between its classification and any further knowledge of it. Wilfred Stalker first collected it on Alexandria Station in the Barkly Tableland at the start of the twentieth century when he was gathering specimens for the British Museum. One of the museum's scientists, George Alfred Boulenger, duly named it Ingram's brown snake after Stalker's employer, Sir William Ingram. But Boulenger simply provided a physical description of the snake, unaware it preys on the rat, which was booming when Stalker collected the snake. Stephen Phillips, a postgraduate student at the University of New England, finally recognised this connection when the rat irrupted in the Northern Territory and Queensland in the early 1990s following a strong La Niña at the end of the 1980s.

Ethabuka Station—a terrain of sand dunes and claypans on the edge of the Simpson Desert in south-western Queensland—was a new site for study of the rat during this plague. Martin Predavec, a postgraduate student at Sydney University, and his supervisor, Chris Dickman, found that when the rats began appearing on Ethabuka, many were juveniles, primarily males. While some of their burrows had as many as fourteen entrances and twenty metres of underground tunnels, others were short with just two entrances. When Predavec and Dickman conducted the only study of the movements of the rat using new technology, fitting tiny radio-transmitters to a few rats and dusting several more with fluorescent pigment, they found that, rather than being confined to one burrow, individual rats used two to three, spending most of their time in them even at night. Almost all their excursions occurred before the moon rose and letter-winged kites could get a better sight of them.

The rat was also the subject of more study on Davenport Downs, where Sidney Jackson encountered it in 1918. A catalyst for this new work was the bilby. Its lesser species became extinct in the 1950s. Its greater species—once found across more than 80 per cent of Australia, including back of the Lachlan where Kenric Harold Bennett encountered it—disappeared from New South Wales before World War I. But tiny populations survived in

Western Australia and Queensland. The biggest Queensland colony was on Davenport Downs where the ground was pock-marked with the burrows of both the rat and the greater bilby, which increasingly was called just the bilby.

Peter McRae, a zoologist with Queensland's National Park and Wildlife Service, best known as one of the 'bilby brothers' at the forefront of trying to preserve it, studied the rat there. He observed that the rains that led to the rat being found on Ethabuka Station triggered a much bigger increase on Davenport Downs where, once again, there were prodigious fluctuations in its numbers in successive years. It peaked in August 1991, collapsed, then peaked again in August 1992 and again plummeted, with McRae providing an unprecedented account of what happened when the rat was under stress at the height of each plague. With their numbers about to crash, the rats were 'in a poor state of health, often blind, displaying cannibalism, active long before and after sunset and sunrise, and were observed drinking water from a bore drain where the estimated water temperatures were above 50° centigrade'.

Davenport Downs was also a research site for Jack Pettigrew of the University of Queensland, whose initial interest was the vision of the letter-winged kite. He found that its adaptation to nocturnal conditions is small but significant and very recent in evolutionary terms, occurring 'as little as 100,000 years ago', so it requires moonlight for efficient hunting. But Pettigrew also studied the impact of the cat and dingo on the bilby and, in July 1992, was horrified when a spotlight at dusk revealed the reflections of a pair of cat's eyes in virtually every coolibah, and nine pairs in a tree where Pettigrew had tagged twenty-six kite chicks a year before. The cats' prime prey was the rat but, because of his concern for the kite and the bilby, Pettigrew and his assistants shot the cats for three nights, killing 182, all sitting in kite nests. Pettigrew also informed Brisbane's *Courier-Mail*, which put this 'environmental tragedy' on its front-page with two photographs by Pettigrew: one, taken on a previous trip, of a kite's nest filled with chicks, and a new one of two cats in a nest.

The federal government had just recognised feral cats as a 'key threat' to the Australian environment. It responded by bringing in the Australian army—perhaps the first and only time it has been deployed for conservationist ends. In three nights, ten snipers killed 423 cats. The largest was seventeen pounds or nearly eight kilograms—almost twice the size of the biggest recorded feral cats, except for one study from western New South Wales. Within a few weeks, Pettigrew's team and members of Queensland's National Parks and Wildlife Service shot 200 more. Sometimes they killed six to eight cats in kite nests in a tree, then killed a similar number the next day in the same tree.

The bilby was fortunate that colonists had adopted its Yuwaalaraay name. The enduring use of the term 'rat' for many other small native species deprived them of public interest and sympathy when they were in jeopardy. In a report published by the federal Nature Conservation Agency in 1995, four leading ecologists and zoologists proposed that Aboriginal names for these animals be adopted. Dick Braithwaite, Steve Morton, Andrew Burbidge and John Calaby called for this change to avoid the connotations of 'introduced vermin,...disease, filth and unscrupulous cunning', and because Aboriginal names were 'far more attractive' and 'belonged to the animals in a much more profound sense' than those currently in use. Given the range of the long-haired rat, the scientists had to select a regional name for national adoption. Not surprisingly, they chose *mayaroo*, as they spelled it. But while their report was brought to a wider audience through magazines such as *New Scientist*, it had no impact.

The decade from 1997 has been dubbed the 'Millennium Drought' because south-eastern Australia was exceptionally dry. But monsoonal rains in the north were very heavy, so the total rainfall for Australia was higher than usual. In the middle of the continent, at Dalhousie Springs, rainfall was about average. While an earlier survey there had found just a few teeth of the long-haired rat, scientists in 2003 found so many rats that they concluded the springs were one of the rat's refuges. Its preference was for areas with date palms, first planted in the 1890s. But because the scientists considered these palms an alien infestation, smothering native growth, they set about removing them.

A very different discovery came from another geological era—perhaps the end of the Pleistocene, perhaps the start of the Holocene. On the western margins of Mygaroo Lakes, south of Alice Springs in the Northern Territory, scientists conducting a botanical survey found fossilised bones on the surface of dunes, prompting excavations at two sites. Among the bones of twelve different vertebrates identifiable with modern species were six lower jaws of the long-haired rat from about 12,000 years ago, its oldest known physical remains.

Another irruption began at the end of the Millennium Drought. The first report came from western Queensland in 2007 following flooding rains at the beginning of the year. There were further reports in 2009 after more big rains and, despite a brief El Niño in between, these localised increases grew due to a particularly strong La Niña starting in 2010, Australia's third wettest year. That April the rat was outside Winton. In May it was just east of the Simpson Desert. In October some areas where it was abundant on the Diamantina and Cooper resembled a 'ploughed paddock'. When one of the great polymaths of the interior, Dick Kimber, found one freshly killed in a wedge-tailed eagle's nest in the southern Simpson Desert, and observed many rat tracks in the nearby sandhill country, he predicted 'a plague of them before the year is out'.

The irruption peaked in 2011, Australia's second-wettest year on record, as the intense La Niña continued and Lake Eyre again partially filled. That January long-haired rats were outside Boulia. In March they began coming through the Lochern National Park south-west of Longreach, while Bedourie was 'all but overtaken'. In April the rats reached Marree in South Australia for the first time since the mid-1970s and were in Alice Springs in the Northern Territory for the first time since 1996. In May they were in sand dunes on the east of the Simpson Desert, suggesting a gradual drift along drainage lines, and near the top of the Birdsville Track in South Australia. In June, they were found in big numbers on Clifton Hills Station where, to the irritation of ecologists seeking to catch the marsupial kowari, the rats monopolised the scientists' traps on the gibber plains. From August until October, they were also on the plains around the Old Andado homestead on the north-western edge of the Simpson Desert in the Northern Territory.

Their taste for soft-coated wiring resulted in the rats destroying the electrics in cars, damaging the computer cables in a police station, and disrupting phone and internet services. They excited most attention when they went off with the false teeth of a resident of Stonehenge in central western Queensland. But there was no panic. The rats' numbers were small, especially given the duration and intensity of this wet period. Their response to people was also less aggressive. A visitor to Old Andado described them as 'very shy' after just a few came onto her swag in the night. A conservation officer advised they were likely to be 'really docile' and would attack only if handled. 'If you pick one up, it might bite,' he instructed.

Legislation, making it an offence to take or kill any native fauna including the rat, had afforded it little or no protection. Lack of enforcement, commonplace with environmental laws, was one problem. Government's power to waive the rat's protection by declaring it an agricultural pest was another. The Western Australian Government did so in 1983–84 when the rat irrupted in the Ord River irrigation area and fed on its rockmelons, maize and sunflowers. Agricultural Protection Board officers in Western Australia, who initially did not know the identity of the rat, took pride in how, by baiting it with grain containing the poison 1080, best known as a rabbit-killer, and by destroying its burrows, they contained the rat when it seemed it could 'be the final straw for struggling famers'. One of the board's officers found that the average size of the rats' litters was more than twelve offspring, the highest figure ever recorded in a particular area.

When the rat irrupted following the Millennium Drought, rural residents typically ignored its protected status. Although officials in the Northern Territory warned that anyone who deliberately killed the rat there could be fined $65,000, many householders baited it. Noonbah Station—a 50,000-hectare cattle property outside Longreach in Queensland, home to Angus and Kate Emmott—was different. A latter-day bush naturalist and fine nature photographer, with much in common with Kenric Harold Bennett, Angus Emmott has created a herbarium of the flora of Noonbah and helped form an entomological collection for south-west Queensland. He has co-authored field guides to snakes and frogs and has two species and two sub-species named after him. A powerful advocate of environmental protection, he has been at the forefront of rural concern about climate change. When the rat appeared on Noonbah in 2011, the Emmotts displayed unprecedented regard for it. As thirty to forty a night entered their homestead, they trapped, then relocated them.

The only area where the rat was the subject of academic study was Ethabuka Station on the edge of the Simpson Desert—long the research ground of Chris Dickman of Sydney University and his colleagues and students. It was a cattle station when Dickman first worked there, but in 2006 it became one of many properties acquired through private philanthropy for environmental management by Australia's Bush Heritage Trust. For all Ethabuka's value as a research site, it had always been marginal to the rat's terrain. The research there in 2011 confirmed that flooding rains were drivers of plagues. It also suggested, contrary to Predavec's and Dickman's work from the early 1990s, that adult rats were at the forefront of the rat's movements, most likely because they had greatest capacity to travel, and there was no differential between the sexes.

The rat also proliferated on Kalamurina, an old cattle station stretching from the Simpson Desert to the northern shore of Lake Eyre, which the Australian Wildlife Conservancy acquired in 2007 for environmental management. The rat was at Kalamurina in 'huge numbers' at the end of 2010. It still 'abounded' at the end of 2011, though its 'numbers dropped off with increasing distance from the water'. In 2012, artist Keith Armstrong recorded there was still 'gross abundance throughout every kind of environment—riverine, far from water, gibber, dry, shrubbed. Everywhere you look there are burrows—feet collapsing into the dune due to their runways.'

National parks were rare in western Queensland until the early 1990s when the government set about doubling their area across the state, with a particular focus on the west. Because of opposition from rural interests, the government failed to make a park on the Bulloo, long recognised as an area of exceptional environmental richness, just as the New South Wales Government did nothing to protect the Bulloo Overflow despite it being another area of great biodiversity. But because of the bilby, the Queensland Government displayed more resolve in relation to Davenport Downs. In 1996, 1740 square kilometres of it became a national park, called 'Astrebla' after the Latin name for the Mitchell tussock grass that grew in the area.

This park was among the areas where the rat boomed following the Millennium Drought. In the winter of 2007, the carnivorous kowari was eating it in small numbers. In late 2009, 'hundreds of thousands' of rats were there. 'Almost every square metre was home to a rat burrow', with some even jumping into the laps of park staff while they ate dinner. Before long, they were occupying empty bilby burrows and feasting on the endangered native melon. When another influx of cats followed, the cats primarily ate the rats—carnage which excited no public interest. But when the rats' irruption ended in 2013 and the cats turned, as usual, to bilbies, which had become increasingly iconic, the government again had some of the cats shot. One of those killed was the stuff of what colonists would have thought of as a 'tough yarn'. With a dozen baby bilbies in its stomach, it was said to weigh fifteen kilograms. Queensland's Minister for National Parks dubbed it a 'miniature tiger'.

If more evidence of the cat's destructiveness were needed, an experiment in the Northern Territory involving the long-haired rat provided it. This experiment was conducted within another private nature reserve—the old Wongalara cattle station on the southern edge of Arnhem Land, which the Australian Wildlife Conservancy had also acquired. Some long-haired rats, specially bred in captivity, were placed in two high-fenced enclosures inaccessible to cats; others were placed in two low-fenced enclosures accessible to cats. Both kinds of enclosure offered dense, diverse vegetation, good for the rats. They survived in the predator-proof enclosures. All were dead in one of the accessible enclosures in a year and a half. In the other, they died in two months, possibly the work of just one or two cats.

EPILOGUE

In the far north-east of the Torres Strait, at the northern tip of the Great Barrier Reef, lies Bramble Cay. A tiny island made up of reef rubble and sand with its highest point just three metres above high tide, its area of vegetation in 1998 was two-and-a-half hectares. By 2011, it was little more than a hectare as sea levels in Torres Strait rose, more than twice the global average, due to climate change. By 2014, it was at most one-tenth of a hectare. During this period, a small rodent with a distinctive mosaic tail, the one mammal unique to the reef, which had been found only on Bramble Cay, disappeared. Most likely, the last *Melomys rubicola* died when the cay was inundated by seawater during storm surges.

Its loss—feared by scientists after limited surveys of the cay in December 2011 and March 2014 did not find any melomys—was confirmed by a comprehensive survey in August–September 2014.

In 2017, members of the Queensland Government's Threatened Species Unit and researchers from the University of Queensland identified Bramble Cay—and hence Australia—as the site of the first mammal extinction on the planet caused primarily if not solely by climate change. That year the Queensland Government declared *Melomys rubicola* extinct. At the start of 2019, the Australian Government followed suit.

John Woinarski, who is at the forefront of studying Australia's extinctions, focused on another cause. Government had failed to move some of the melomys and breed them in captivity while there was time. It could have done so when a census of the cay in 1998 suggested there were fewer than a hundred melomys. It might still have done so after a survey in 2002 found only ten. As part of identifying the melomys's extinction as preventable, Woinarski blamed the Queensland environment department, 'or one or two key personnel within it', who were 'unsympathetic to captive breeding generally' and did nothing for the melomys, despite its manifest vulnerability.

The federal government was also to blame because of its national responsibility for Australia's endangered species. All it had done was approve a 'recovery plan' for the melomys—never implemented in any respect—which suggested in 2008 that the 'likely consequences of climate change, including sea-level rise and increase in the frequency and intensity of tropical storms', were 'unlikely to have any impact' on the melomys over the next five years. The melomys was also on a remote, uninhabited island. It was not a charismatic species, with great popular appeal, as the bilby had become. Rather, its name worked against it. 'Intervention was impeded by the perception that this was a rat', zoologist Graham Fulton observed.

Australia already had by far the worst record for loss of

mammals of any continent. Only the sea mink has become extinct in North America since 1500, but the melomys is the thirtieth Australian species to die out since 1788. While about 1.5 per cent of mammals across the planet have been lost since 1500, the proportion in Australia since 1788 is higher than ten per cent. Big animals have proved most vulnerable elsewhere, but smaller ones, weighing between 35 grams and 5.5 kilograms, have been worst affected in Australia. To read the list of Australia's extinct species is to go from one 'rat' or 'mouse' or 'mys' to another—fifteen in all—with the 'rats' ranging from the oolacunta, or desert rat-kangaroo, to the lesser stick-nest rat, the white-footed rabbit rat, and two forms of *Rattus*, the bulldog rat and Maclear's rat.

The long-haired rat, though within the weight range of Australian mammals at particular risk, is not in immediate jeopardy. On the 'Red List' of threatened species compiled by the International Union for the Conservation of Nature, it is a species of 'least concern'—not requiring special measures or protections, unlike those categorised as vulnerable, endangered and critically endangered. The IUCN's assessment of the rat, led by John Woinarski, recognises that there are far fewer irruptions of the rat now than in the past, perhaps due to degradation of the habitat of its refuges. The assessment also recognises that the rat 'is mainly restricted to core areas and is more susceptible to threats during drought years'. But it suggests the rat still 'has a relatively wide distribution, presumed large population, and it is unlikely to be declining at nearly the rate required to qualify for listing in a threatened category'.

Much of this assessment is clear-cut. Since the mid-1970s, the gaps between the rat's major irruptions have grown markedly, but the rat still has refuges in the Barkly Tableland and the Channel

Country, as well as elsewhere. Some are longstanding such as the Bulloo Overflow. Others have been created relatively recently around bores and irrigation areas, including within the Ord River Scheme. In *The Complete Guide to Finding Australian Mammals*, David Andrews provides instructions about how to reach one colony outside Kununurra. 'Look for them in canefields near town,' he advises. 'Take Ivanhoe Rd, right on to Research Station Rd, bear left at a Y junction at the end and check canefields around there.'

Yet, as Woinarski has discussed, there is a mismatch between how the IUCN assesses the decline of animals and what has happened in Australia. The international body is concerned only with a decline over a ten-year period, unless a species has been reduced to tiny numbers or is found in just a small area. But many Australian species have declined radically since 1788 and are still doing so. Undiscussed by scientists, whose studies of the long-haired rat have typically focused on small areas over short periods, the plagues of the rat since the mid-1970s have been relatively modest. The talk of hordes, myriads, legions, armies and millions has almost stopped. While there are no precise figures, the long-haired rat has experienced a dramatic collapse in its abundance.

The causes of the decline of species are not always easy to identify, even on a general level, and the relative importance of each cause, when more than one is involved, is usually much harder to determine. When Gerard Krefft provided the first substantial account of the impact of European colonisation on Australia's mammals after his stint near the junction of the Murray and Darling rivers in the late 1850s, he blamed the settlers' livestock for the disappearance of five species. Some later scientists have questioned Krefft's explanation as simplistic, suggesting changed fire regimes, clearing of habitat, and cats and foxes would have contributed. Other scientists have supported Krefft's analysis.

While bigger monsoonal rains in Australia's north caused by climate change could benefit the long-haired rat, very heavy rains can also cause its numbers to plummet as began to become clear in the early 1940s. Further south, more frequent, more intense droughts are increasing the rat's jeopardy through reduced water and diminished plant growth and cover, facilitating predation by cats and foxes. Some of the places that have served as refuges for the rat, allowing it to survive in low numbers through dry years, may have already failed, without any awareness of what has occurred because these refuges have never been located or are only infrequently studied. Other refuges may fail soon. All may fail before the end of this century.

The letter-winged kite is one indicator. As recently as 1993, Penny Olsen, one of Australia's authorities on raptors, observed that when the rats are abundant, the kite also breeds unusually fast, 'raising several broods of larger clutches in a season, and the juveniles may also breed within weeks of leaving the nest'. She elaborated: 'Groups of kites follow the course of the rats, breeding on waterholes in colonies with as many as fifty nests.' When drought returns, 'a core of kites remains inland', surviving on other food including insects, but the remainder 'is forced to leave or starve', sometimes reaching the coast. While the kite appears to be 'a completely nocturnal hunter...when rats are plaguing and food is plentiful...many observers have seen the kite on the wing during the day and even hunting during daylight'.

Much of this account no longer holds. While pellets below the kites' nests in western Queensland provided early evidence of the rat's numbers increasing in 2007, the irruption that followed was notable for the few sightings of the kites inland where their numbers at some of their usual roosting sites were down by more than ninety per cent. What occurred after the rat plague was also striking. There was just one set of reports, from South Australia, of the kite reaching the coast. While more kites have been observed since, Australia's State of the Environment Report for 2016 expressed 'grave concerns' for the kite's 'continued persistence'. When ornithologists and ecologists sought to explain its jeopardy, they fixed on how cats both compete for prey with the kite and prey on it. They might also have considered the rat's decline. If not even the exceptional rains of 2010 and 2011 could lead the rat to irrupt on an immense scale, what will?

The *mayaroo*'s great days and those of its kite are gone.

SOURCES

This book draws on a wide array of material. It includes Kenric Harold Bennett's 68 letters to Edward Pierson Ramsay, which are part of the Ramsay correspondence papers in the State Library of New South Wales at ML MSS1589, Bennett correspondence in the Australian Museum at AMS 8, and the Museum's schedules for purchases from Bennett at AMS 55. A selection of key sources follows, primarily for the long-haired rat.

INTRODUCTION

The first rendering of the letter-winged kite is 'Watling Drawing—no 99' in the Natural History Museum, London, discussed by K. A. Hindwood, 'The Watling Drawings', *Proceedings of the Royal Zoological Society of New South Wales*, 1968–9, pp. 16–32. On boom and bust, see, especially, S. R. Morton *et al*, 'A Fresh Framework for the Ecology of Arid Australia', *Journal of Arid Environments*, 75, 2011, pp. 313 28; Richard Jordan *et al*, 'Boom and Bust (or not?) among Birds in an Australian Semi-desert', *Journal of Arid Environments*, 139, 2017, pp. 58–66. For the long history of the rat, see J. H. Robins *et al*, 'Evolutionary Relationships and Divergence Times among the Native Rats of Australia', *BMC Evolutionary Biology*, 10, 2010, p. 375. For European responses to rats, see Jonathan Burt, *Rat*, Reaktion Books, London, 2006. Jonathan Rosen wrote about 'forensic ornithology' in 'The Birds', *New Yorker*, 6 January 2014. For the 'rat years', see *North Queensland Register*, 24 July 1905, p. 63. For Bennett and bushrangers, see *Gippsland Guardian*, 16 December 1867, p. 3. For his part in a conflict between squatters and a selector, see *Burrowa News*, 4 November 1881, p. 2; 2 December 1881, p. 2; 27 January 1882, p. 3. Tim Low's article is 'Why We Need a Better Appreciation of Australia's Rodents', *Australian Geographic*, 21 March 2018.

CHAPTER 1
A Word from the Barngarla

The best discussion of *mai erri*, identifying it as the long-haired rat, is in Dorothy Tunbridge, *The Story of the Flinders Ranges Mammals*, Kangaroo Press, Kenthurst, 1991. See also T.H. Johnston, 'Aboriginal Names and Utilization of the Fauna in the Eyrean Region', *Transactions of the Royal Society of South Australia*, 67, 1945, pp. 244–311; Peter K. Austin, *A Grammar of Diyari, South Australia*, Cambridge University Press, Cambridge, 1981. Schürmann wrote *Vocabulary of the Parnkalla Language spoken by the Natives inhabiting the Western Shores of Spencer's Gulf*, Dehane, Adelaide, 1844. See, further, Mark Clendon, *Clamor Schürmann's Barngarla Grammar*, University of Adelaide Press, Adelaide, 2015; Kim McCaul, 'Clamor Schürmann's Contribution to the Ethnographic Record for Eyre Peninsula, South Australia', in Nicolas Peterson & Anna Kenny (eds), *German Ethnography in Australia*, ANU Press, Canberra, 2017. On word lists, see Luise Hercus, 'The Marawara Language of Yelta: Interpreting Linguistic Records of the Past', *Aboriginal History*, 8(1), 1984, pp. 56–62.

CHAPTER 2
The Blunders of Science

For Kennedy's expedition and Thomas Wall, see Edgar Beale, *The Barcoo and Beyond 1847*, Blubber Head, Hobart, 1983; John Gould, *Mammals of Australia*, Gould, London, 1845–1863, part 6; J.H. Calaby & J.M. Taylor, 'Type Locality of *Rattus villosissimus*', *Australian Mammalogy*, 1, 1974, pp. 267–8. Gray published the rat in his *List of the Specimens of Mammalia in the Collection of the British Museum*, British Museum, London, 1843. See, further, J. Mary Taylor & B. Elizabeth Horner, 'Results of the Archbold Expeditions: No 98 Systematics of

Native Australian *Rattus*', *Bulletin of the American Museum of Natural History*, 150, 1973, pp. 1–130.

CHAPTER 3

A Wonder of Art

This chapter depends largely on a reconsideration of standard sources: *Report of the Commissioners appointed to Enquire into and Report upon the Circumstances connected with the Sufferings and Death of Robert O'Hara Burke and William John Wills*, Ferres, Melbourne, 1862; Marjorie Tipping, *Ludwig Becker: Artist and Naturalist with the Burke and Wills Expedition*, Melbourne University Press, Melbourne, 1979; Hermann Beckler, *A Journey to Cooper's Creek*, Melbourne University Press, Melbourne, 1993. For Burke's response to the rat, see his dispatch, 13 December 1860, VEE papers, 2082/1a(13), SLV; and for Wright's rat-killing, his diary, 21 April 1861, VEE papers, 2083/3(b), SLV. James Rodwell wrote *The Rat: Its History and Destructive Character*, Routledge, London, 1858. Edwin Welch's account is in 'The Story of Burke and Wills', ML A1928, p. 54, SLNSW. See, also, my *Burke and Wills: From Melbourne to Myth*, David Ell, Sydney, 1991; E. B. Joyce & D. A. McCann (eds), *Burke and Wills: The Scientific Legacy of the Victorian Exploring Expedition*, CSIRO, Melbourne, 2011; Ian D. Clark & Fred Cahir (eds), *The Aboriginal Story of Burke and Wills*, CSIRO, Melbourne, 2016. Barrett Reid published my first piece about Becker's watercolour, 'A Rat's Tale', in *Overland*, 127, 1992, pp. 7–9.

CHAPTER 4

Bush Naturalists

For Price Fletcher and Charles Darwin, see *Queenslander*, 14 December 1878, p. 20; for Fletcher on environmental destruction, see *Queenslander*, 4 May 1878, p. 148; more generally, 24 February 1906, p. 8; the Egyptian plague, 21 December 1878, p. 365. For Feilberg, see *Queenslander*, 1 May 1880, p. 560. For the first response to the rat, see *Queenslander*, 8 March 1879, p. 301. For W. A. Brodribb, see his *Recollections of an Australian Squatter*, Woods, Sydney, 1883. For

Moolbong and Yandembah, see *Mackenzie's Riverina: A Tour of the Hay District Pastoral Holdings in the 1890s*, Hay Historical Society, Hay, 2002. For Yandembah's pine, see *Australian Town and Country Journal*, 6 August 1881, p. 19; and the station's refinement, *Riverine Grazier*, 5 September 1877, p. 5. On Moolah, see *Riverine Grazier*, 6 June 1877, p. 2. For Bennett's first dealings with Price Fletcher and his pet kangaroo, kookaburras and mallee hens, see *Queenslander*, 22 February 1879, p. 245; 21 June 1879, p. 780; 20 November 1880, p. 653; 15 January 1881, p. 80. For Thickthorn, see *Queenslander*, 18 March 1882, p. 332. For Bennett's long letters, see *Queenslander*, 14 June 1879, p. 746; 26 July 1879, p. 107; and for a response, 30 August 1879, p. 268. On stick-nest rats, see P. Copley, 'Natural Histories of Australia's Stick-nest Rats', *Wildlife Research*, 26, 1999, pp. 513–39. On Krefft, see his *Two Papers on the Vertebrata of the Lower Murray and Darling*, Reading & Wellbank, Sydney 1865; P. W. Menkhorst, 'Blandowski's Mammals: Clues to a Lost World', *Proceedings of the Royal Society of Victoria*, 121, 2009, pp. 61–89.

CHAPTER 5

A Perfect Egyptian Plague

On rats during the siege of Paris, linked to Australia, see *Australian Town and Country Journal*, 22 April 1871, p. 6. For the Australian plague in the *Queenslander*, see 21 December 1878, p. 365; 8 February 1879, p. 180; 8 March 1879, p. 301; 3 May 1879, p. 563. On refuges, see Chris R. Pavey *et al*, 'The Role of Refuges in the Persistence of Australian Dryland Mammals', *Biological Reviews*, 92, 2017, pp. 647–64. Edward Palmer discussed his treatment of Aboriginal people in *Brisbane Courier*, 22 July 1874, p. 6; letter to A. W. Howitt, 5 August 1882, Museum of Victoria, XM 276. He wrote 'Notes on a Great Visitation of Rats in the North and North-Western Plain Country of Queensland in 1869 and 1870', *Proceedings of the Royal Society of Queensland*, 2, 1885, pp. 193–8. For other reports of the rat, see *Brisbane Courier*, 15 April 1871, p. 5; *South Australian Register*, 14 April 1871, p. 5; 21 April 1871, p. 4; 16

November 1871, p. 6; *Evening News*, 27 April 1871, p. 2; 5 May 1871, p. 3; *Express and Telegraph*, 15 June 1871, p. 2; *Northern Argus*, 3 May 1872, p. 2; 23 August 1872, p. 2; *South Australian Register*, 24 April 1873, p. 2. For the Flinders Ranges, see Dorothy Tunbridge's *Flinders Ranges Dreaming*, Aboriginal Studies Press, Canberra, 1988; *Flinders Ranges Dreaming Sites Record*, Aboriginal Studies Press, Canberra, 1991; and *The Story of the Flinders Ranges Mammals*, Kangaroo Press, Kenthurst, 1991; Meredith Smith, 'Remains of Mammals including *Notomys longicaudatus* in Owl Pellets from the Flinders Ranges, S.A.', *Australian Wildlife Research*, 4, 1977, pp. 159–70. More generally, for these pellets, see A.C. Robinson *et al*, 'The Rodents of South Australia', *Wildlife Research*, 27, 2000, pp. 379–404.

CHAPTER 6

The Night Kite

Bennett wrote about the black falcon and the letter-winged kite in 'Notes on the Habits &c of Birds breeding in the Interior of New South Wales', *Proceedings of the Linnean Society of New South Wales*, 1885, pp. 162–9. For Samuel White, see S.A. Parker, 'Samuel White's Ornithological Expeditions in Northern South Australia in 1863', *South Australian Ornithologist*, 28, 1980, pp. 113–9; Sylvester Diggles, see *Queenslander*, 13 March 1875, p. 7. For the bilby and Bennett's contributions to the 1879 exhibition, see *Riverine Grazier*, 6 June 1877, p. 1; *Australian Town and Country Journal*, 15 November 1879, p. 8; 14 February 1880, p. 18; *Queanbeyan Age*, 28 February 1880, p. 3; *Sydney Mail*, 10 April 1880, p. 693. Bennett wrote 'Notes on the Black-breasted Buzzard', *Proceedings of the Linnean Society of New South Wales*, 1881, pp. 146–8; 'Notes on the Method of Obtaining Water from Eucalyptus Roots as practised by the Natives', *Proceedings of the Linnean Society of New South Wales*, 1883, pp. 213–5; and first wrote about the kite in the *Queenslander*, 26 July 1879, p. 106.

CHAPTER 7

City of Iron

For Winton in 1880, see *Telegraph*, Brisbane, 20 July 1880, p. 6; *Queenslander*, 9 October 1880, p. 475. For rats elsewhere, see *Queenslander*, 22 May 1880, p. 658; *South Australian Advertiser*, 8 February 1879, p. 6; 22 July 1880, p. 4; *South Australian Chronicle*, 27 March 1880, p. 7; Edward B. Sanger, 'The Mammalian Fauna of the Australian Desert', *American Naturalist*, 18, January 1884, pp. 9–12. For Bennett and the impact of possums, see his 'Remarks on the Decay of Certain Species of Eucalypti', *Proceedings of the Linnean Society of New South Wales*, 1885, pp. 453–4, and for Bennett and the dingo, see *Riverine Grazier*, 6 June 1877, p. 1; *Queenslander*, 15 January 1881, p. 80. For rats in and around Winton, see *Brisbane Courier*, 26 May 1880, p. 2; 12 June 1880, p. 6; 27 September 1880, p. 3; *Queenslander*, 17 July 1880, p. 3; 24 July 1880, p. 121; *Northern Miner*, 14 October 1880, p. 3; *Telegraph*, Brisbane, 3 February 1883, p. 5; Carl Lumholtz, *Among Cannibals*, Murray, London, 1889; W.H. Corfield, *Reminiscences of Queensland 1862–1899*, Frater, Brisbane, 1921; *Central Queensland Herald*, 19 January 1933, p. 46; *Townsville Daily Bulletin*, 28 November 1945, p. 6. On Price Fletcher's monthly notes, see *Queenslander*, 16 October 1880, p. 493; 20 November 1880, p. 653; 5 February 1881, p. 174.

CHAPTER 8

Year One

For Yandembah in 1884, see *Evening News*, 23 April 1884, p. 7; for Winton, see *Western Champion*, 25 July 1884, p. 2; *Queensland Figaro*, 20 December 1884, p. 23. For the identification of La Niña and El Niño years, see Joëlle Gergis & Linden Ashcroft, 'Rainfall Variations in South-eastern Australia Part 2', *International Journal of Climatology*, 33, 2013, pp. 2973–87; Linden Ashcroft *et al*, 'Southeastern Australian Climate Variability 1860–2009', *International Journal of Climatology*, 34, 2014, pp. 1928–44. For Winton in 1885, see *Morning Bulletin*, 28 January 1885, p. 5; *Brisbane Courier*, 27 February 1885, p. 5; *Capricornian*, 14 March

1885, p. 4. For the rat there, see *Capricornian*, 15 August 1885, p. 24. For rats on the Burke River, see W. H. Davidson, 'Western Experiences in 1885–6', *Queensland Geographical Journal*, 34–35, 1918–20, pp. 43–58; at the Paravituary Waterhole, *Telegraph*, 5 January 1887, p. 2; the private survey party, *Chronicle*, Adelaide, 14 February 1935, p. 48; on David Lindsay's expedition, *South Australian Advertiser*, 2 July 1887, p. 6.

CHAPTER 9

Year Two

For the rat on the Bulloo, see *Evening News*, 20 January 1886, p. 5; *Riverine Grazier*, 3 February 1886, p. 2; *South Australian Advertiser*, 26 March 1886, p. 6; *Brisbane Courier*, 6 April 1886, p. 6; *Globe*, 8 April 1886, p. 6; *Queensland Times*, 8 April 1886, p. 5; at Birdsville, see *South Australian Register*, 7 April 1886, p. 3; *Queenslander*, 24 April 1886, p. 630; the Georgina, see *Brisbane Courier*, 4 May 1886, p. 4; Eyre's Creek, *Queenslander*, 5 June 1886, p. 885; Windorah, *Queensland Figaro*, 7 August 1886, p. 6; Diamantina, *South Australian Register*, 30 December 1886, p. 3; Rabbit Inspector Holding, *Globe*, 1 May 1886, p. 5; Cobham Lake, *Daily Telegraph*, 14 June 1886, p. 3; Silverton and Broken Hill, *Yorke's Peninsula Advertiser*, 6 August 1886, p. 3; *Port Augusta Despatch*, 6 August 1886, p. 3; Hergott Springs, *Adelaide Observer*, 13 February 1886, p. 9; Parallana station, *Express*, 13 March 1886, p. 2; Farina, *South Australian Register*, 2 April 1886, p. 4; Great Northern Railway, *South Australian Advertiser*, 23 April 1886, p. 5; *Evening Journal*, 8 June 1886, p. 2; Camel-carrying Company, *South Australian Weekly Chronicle*, 26 June 1886, p. 10; Strangways Springs, *Port Augusta Despatch*, 21 June 1886, p. 3; Teetulpa, *Bunyip*, 12 November 1886, p. 2. For Bagot's recollections, see J. B. Cleland, 'Previous Phenomenal Visitations of Rats and Mice in Australia', *Journal and Proceedings of the Royal Society of New South Wales*, 52, 1918, pp. 1–165. For the rat's bones in owl pellets on the Nullarbor, see Taylor & Horner, 'Results of the Archbold Expeditions'; Tunbridge, *The Story of the Flinders Ranges Mammals*, Kangaroo Press, Kenthurst, 1991.

CHAPTER 10

Year Three

For rats in Birdsville, see *Brisbane Courier*, 4 October 1887, p. 5; 29 November 1887, p. 5; Warrego and Paroo, *Glen Innes Examiner*, 20 September 1887, p. 3; Barcoo, *Queenslander*, 24 December 1887, p. 1018; the Cooper, *Queenslander*, 3 September 1887, p. 376; Cockburn, *South Australian Register*, 22 April 1887, p. 6; Eleanor Creek, *Port Augusta Despatch*, 1 February 1887, p. 2; Teetulpa, *Argus*, 28 January 1887, p. 9; *Observer*, 30 October 1926, p. 63; Menindee, *Riverine Recorder*, 12 January 1887, p. 3; 9 February 1887, p. 1; Bourke, *Australian Town and Country Journal*, 26 March 1887, p. 11; Hay, *Australian Town and Country Journal*, 1 October 1887, p. 16; Balranald, *Riverine Grazier*, 10 May 1887, p. 2; 20 September 1887, p. 2; *Australian Town and Country Journal*, 8 October 1887, p. 14; Walgett, *Maitland Mercury*, 10 May 1887, p. 5. For the rat's larger course through New South Wales, see *Sydney Mail*, 28 May 1887, p. 1114. On the wedgetail, see *Riverine Grazier*, 9 April 1884, p. 2; *Sydney Morning Herald*, 18 May 1885, p. 4. The catalogue of Bennett's Aboriginal collection is *Descriptive List of Australian Aboriginal Weapons, Implements &c from the Darling and Lachlan Rivers in the Australian Museum*, Potter, Sydney, 1887. Bennett's mapping is in 'Notes on a Species of Rat (*Mus tompsonii*) now infesting the Western Portion of New South Wales', *Proceedings of the Linnean Society of New South Wales*, 1887, pp. 447–9.

CHAPTER 11

Year Four

Neville Nicholls identified 'The Centennial Drought' in Eric Webb (ed.), *Windows on Meteorology: Australian Perspectives*, CSIRO, Melbourne, 1997, pp. 118–26. Charles Todd published his observations in *Argus*, 27 December 1888, p. 10. For the rat in Queensland, see *Queenslander*, 4 August 1888, p. 197; *Western Champion*, 10 April 1888, p. 2; *Capricornian*, 21 April 1888, p. 15; *Northern Standard*, 22 April 1932, p. 7. For the rat in New South Wales, see *Sydney Mail*, 21 January 1888, p. 154;

Bathurst Free Press, 3 March 1888, p. 5. For Victoria, see *Kerang Times*, 7 February 1888, p. 3; 24 April 1888, p. 2; 25 May 1888, p. 4; *Herald*, 14 April 1888, p. 3; *Mount Alexander Mail*, 20 April 1888, p. 2; *Leader*, 21 April 1888, p. 10.

CHAPTER 12
The Name of the Beast

For Humphry Davy on the rat, see *Riverine Grazier*, 12 April 1887, p. 2. For Bulloo Downs, the Upper Darling and the Paroo, see *Telegraph*, Brisbane, 11 September 1880, p. 3; *Adelaide Observer*, 3 December 1887, p. 43; *Australian Town and Country Journal*, 1 October 1887, p. 16. For Kidman, Wheeler and Burston, see *Observer*, Adelaide, 14 July 1928, p. 22; *Brisbane Courier*, 13 February 1929, p. 7. Bennett's published account is 'Notes on a Species of Rat (*Mus tompsonii*) now infesting the Western Portion of New South Wales', *Proceedings of the Linnean Society of New South Wales*, 1887, pp. 447–9. For the confusion caused by the Norway rat, see *Sydney Mail*, 22 February 1873, p. 234. For other descriptions of the rat in 1887, see *Riverine Grazier*, 1 April 1887, p. 2; 19 April 1887, p. 2; *Australian Town and Country Journal*, 9 April 1887, p. 744. For 16 inches, see *Sydney Mail*, 14 August 1886, p. 332. For Ganpung Homestead, see Murray Ellis, 'A Discussion of the Large Extinct Rodents of Mootwingee National Park, Western New South Wales', *Australian Zoologist*, 20, 1995, pp. 1–4.

CHAPTER 13
If Rats Kill Young Cats

For one response to Ranken's climate writing, see *Sydney Morning Herald*, 29 January 1874, p. 6. For the boundary rider's cats on Tongo, see *South Australian Advertiser*, 9 April 1887, p. 7. On Hebden and his cats, see *Brisbane Courier*, 7 October 1885, p. 9; *Australian Town and Country Journal*, 10 October 1885, pp. 22–23; *Sydney Mail*, 2 January 1886, p. 14; *Herald*, 27 March 1886, p. 2; *Australasian*, 14 April 1888, p. 18; *North West Post*, 26 October 1889, p. 3. On Frederick Campbell and the native cat, see *Sydney Morning Herald*, 25 July 1885, p. 8. On the native cats for Tongo, see *Queenslander*, 17 April 1886, p. 630; *Australian Town and Country Journal*, 1 May 1886, p. 23. On the introduced cats for Tolarno, see *Sydney Mail*, 30 October 1886, p. 905; 26 February 1887, p. 437; on those for Thargomindah, see *Australian Town and Country Journal*, 10 March 1888, p. 17; *Brisbane Courier*, 10 July 1888, p. 5; and Barenya, *Townsville Daily Bulletin*, 23 October 1934, p. 9. For the rat as a rabbit-killer, see *South Australian Advertiser*, 20 February 1886, p. 19; *South Australian Register*, 17 May 1886, p. 2; *Sydney Mail*, 8 May 1886, p. 949; 10 August 1886, p. 5; *Queenslander*, 21 August 1886, p. 306; *Australian Town and Country Journal*, 9 April 1887, p. 16; *Riverine Grazier*, 12 April 1887, p. 2. On the rat swimming, see *Sydney Mail*, 28 May 1887, p. 1114, and the rat transforming, see *Riverine Grazier*, 20 September 1887, p. 2; *Australian Town and Country Journal*, 1 October 1887, p. 16. For Kingsmill, see *Evening News*, 12 September 1887, p. 7.

CHAPTER 14
Hatred and Suspicion

For the bilby as a hybrid, see *Freeman's Journal*, 7 April 1900, p. 23. For cat rabbits, see *Sydney Morning Herald*, 31 August 1886, p. 5; *Australasian Sketcher*, 23 February 1888, p. 23; *Argus*, 22 March 1888, p. 5. For the bunyip, see William Ranken, *The Dominion of Australia*, Chapman & Hall, London, 1874; *Evening News*, 14 January 1885, p. 6; *Riverine Grazier*, 16 November 1886, p. 2; 12 April 1887, p. 2. For disparagement of rabbiters, see *Freeman's Journal*, 18 June 1887, p. 13; inspectors, see *Daily Telegraph*, 7 November 1887, p. 5; pastoralists, see *Sydney Morning Herald*, 26 August 1887, p. 4. For the purchases of Aboriginal rabbiters, see *Newcastle Morning Herald*, 14 January 1887, p. 7; of other rabbiters, *Daily Telegraph*, 24 August 1887, p. 5; rabbiter violence, *Riverine Grazier*, 17 May 1887, p. 2; the Tilpa killing, *Australian Star*, 15 February 1888, p. 6. For kangaroo trappers, see *Cootamundra Herald*, 24 March 1886, p. 3; for the Weinteriga case, *Australian Town and Country Journal*, 5 February 1887, p. 4; *Sydney Mail*, 5 February

1887, p. 272; 4 June 1887, p. 1190; *Riverine Grazier*, 15 March 1887, p. 4; *Daily Telegraph*, 30 May 1887, p. 7; for the Gol Gol case, see *Riverine Recorder*, 1 June 1887, p. 2; 23 November 1887, p. 2; *Wilcannia Times*, 2 December 1887, p. 4.

CHAPTER 15
On the Shores of Lake Killalpaninna

This chapter draws primarily on Reuther's ethnography, *The Diyari*, most readily accessible through the translation by Philipp A. Scherer, published by the Australian Institute of Aboriginal Studies in 1981, available on microfiche. For context, see Philip Jones & Peter Sutton, *Art and Land*, South Australian Museum, Adelaide, 1986; L. Hercus & V. Potezny, 'Locating Aboriginal Sites: A Note on J.G. Reuther and the Hillier Map of 1904', *Records of the South Australian Museum*, 24, 1987, pp. 139–51; Isabel McBryde, 'Goods from Another Country: Exchange Networks and the People of the Lake Eyre Basin', in D. J. Mulvaney & J. Peter White (eds), *Australians to 1788*, Fairfax, Syme & Weldon, Sydney, 1987, pp. 253–73; Christine Stevens, *White Man's Dreaming: Killalpaninna Mission 1866–1915*, Oxford University Press, Melbourne, 1994; Philip Jones, 'Naming the Dead Heart: Hillier's Map and Reuther's Gazetteer of 2468 Placenames in North-eastern South Australia', in Luise Hercus *et al*, *The Land is a Map: Placenames of Indigenous Origin in Australia*, Pandanus, Canberra, 2002, pp. 157–73; Jeremy Beckett & Luise Hercus, *Two Rainbow Serpents Travelling: Mura Track narratives from the 'Corner Country'*, ANU Press, Canberra, 2009; Luise Hercus, 'Archaisms in Placenames in Arabana and Wangkangurru Country', in Robert Mailhammer (ed.), *Lexical and Structural Etymology*, de Gruyter, Berlin, 2013; Rod Lucas & Dean Fergie, 'Pulcaracuranie: Losing and Finding a Cosmic Centre with the Help of J. G. Reuther and Others' and Luise Hercus, 'Looking at some Details of Reuther's Work' in Nicolas Peterson & Anna Kenny (eds), *German Ethnography in Australia*, ANU Press, Canberra, 2017.

CHAPTER 16
Villosissimus

North's book is *Descriptive Catalogue of the Nests and Eggs of Birds Found Breeding in Australia and Tasmania*, White, Sydney, 1889. For the eastern-hare wallaby, see Tim Flannery & Peter Schouten, *A Gap in Nature: Discovering the World's Extinct Animals*, Text Publishing, Melbourne, 2001. For Bennett on water conservation, see *Queenslander*, 8 October 1892, p. 692. His last paper on the disappearance of birds is in *Records of the Australian Museum*, 1, 1890–1, pp. 107–9; *Australasian*, 25 April 1891, p. 11. For his death, see *Proceedings of the Linnean Society of New South Wales*, 1891; *Sydney Mail*, 26 March 1892, p. 682. For *Corvus bennetti*, see North's *Nests and Eggs of Birds Found Breeding in Australia and Tasmania*, White, Sydney, 1901, vol 1. For Baldwin Spencer and his collecting see, John Mulvaney *et al*, *'My Dear Spencer': The Letters of F. J. Gillen to Baldwin Spencer*, Hyland House, Melbourne, 1997; John Mulvaney, *From the Frontier: Outback Letters to Baldwin Spencer*, Allen & Unwin, Sydney, 2000. Edgar Waite reclassified the rat in 'Observations on Muridae from Central Australia', *Proceedings of the Royal Society of Victoria*, 1898, pp. 114–28. For the rat in the early 1900s, see *Northern Miner*, 29 June 1900, p. 3; *Northern Territory Times*, 14 July 1905, p. 3. For Stalker's employment, see *Queenslander*, 19 March 1910, p. 37. He wrote about the rat in letters to Oldfield Thomas in 1905–6, available on MC2597 of the Australian Joint Copying Project. Oldfield Thomas did so in 'On Mammals from Northern Australia presented to the National Museum by Sir Wm. Ingram, Bt. and the Hon. John Forrest', *Proceedings of the Zoological Society of London*, 1906, pp. 536–43; and 'On Three New Australian Rats', *Annals and Magazine of Natural History*, 8, 1921, pp. 618–22. For the plague, see *Examiner*, 29 August 1903, p. 3. Heber Longman wrote *Notes on Classification of Common Rodents*, Mullet, Melbourne, 1916. Louis de Rougemont's account of the rat appeared in the *Wide World Magazine*, 2 (11), 1899, pp. 531–42. Conrick's and Archer's reminiscences are in *News*, 13 August 1923, p. 9; *Argus*, 6 May 1939, p. 14. Sidney Pearson wrote for *Hamilton Spectator*, 4 June 1904, p. 2.

CHAPTER 17
Three Cheers for the Diamantina

Jackson's diary for the Diamantina in 1918 is in MS 466 in the NLA. White wrote about the expedition in 'The Letter-winged Kite (*Elanus scriptus*, Gould)', *Emu*, 18, 1919, pp. 157–9; and Jackson in 'Haunts of the Letter-winged Kite (*Elanus scriptus*, Gould)', *Emu*, 18, 1919, pp. 160–72; also in *World's News*, 21 Augusts 1935, p. 10. For the rat in 1908–10, see *Capricornian*, 7 November 1908, p. 26; 5 June 1909, p. 39; *Morning Bulletin*, 11 October 1910, p. 4. For the mouse and the rat in 1918, see *Telegraph*, 1 June 1918, p. 10; *Warwick Examiner*, 22 July 1918, p. 3; 24 September 1919, p. 3; *Gympie Times*, 17 August 1918, p. 6; *Daily Mail*, 8 May 1920, p. 9. Cleland's paper is 'Previous Phenomenal Visitations of Rats and Mice in Australia', *Journal and Proceedings of the Royal Society of New South Wales*, 52, 1918, pp. 1–165. Le Souef's and Burrell's book is *The Wild Animals of Australasia*, Harrap, London, 1926. See also Judy White, *Sidney William Jackson: Bush Photographer 1873 to 1946*, Seven Press, Scone, 1991.

CHAPTER 18
The Summer Vacationist

For Finlayson and the oolacunta, see *Advertiser*, 26 March 1932, p. 15; H. H. Finlayson, 'Rediscovery of *Caloprymnus campestris*', *Nature*, 129, 1932, p. 871; '*Caloprymnus campestris*: Its Recurrence and Characters', *Transactions and Proceedings of the Royal Society of South Australia*, 56, 1932, pp. 148–167; *The Red Centre*, Angus & Robertson, Sydney, 1935; Don Tonkin, *A Truly Remarkable Man: H. H. Finlayson and his Adventures*, Seaview, Henley Beach, 2001. For Jimmy Naylon, see Luise Hercus, 'Leaving the Simpson', *Aboriginal History*, 9, 1985, pp. 22–43; Luise Hercus & Peter Clarke, 'Nine Simpson Desert Wells', *Archaeology in Oceania*, 21, 1986, pp. 51–62; Philip Jones, 'Ngapamanha: A Case Study in the Population History of North-eastern South Australia', in P. Austin *et al* (ed.), *Language and History: Essays in Honour of Luise A. Hercus*, Pacific Linguistics, Canberra, 1991. For Finlayson and the rat, see his 'On Mammals from the Lake Eyre Basin, Part IV', 'On Mammals from the Lake Eyre Basin, Part V', *Transactions of the Royal Society of South Australia*, 63, 1939, pp. 88–117, 348–53. For the rat at Winton, see *Northern Miner*, 16 December 1931, p. 6. For fish eating the rat, see *Cloncurry Advocate*, 31 January 1931, p. 8; *Sunday Mail*, 30 August 1931, p. 13; Erin Kelly *et al*, 'Mammal Predation by an Ariid Catfish in a Dryland River in Western Australia', *Journal of Arid Environments*, 135, 2016, pp. 9–11. For Longreach, Julia Creek, Rankin River and Barkly Tableland, see *Longreach Leader*, 17 April 1931, p. 12; *Sydney Mail*, 3 May 1933, p. 36; *Townsville Daily Bulletin*, 1 September 1934, p. 7; 21 February 1935, p. 12; 1 May 1941, p. 3. For Idriess, see *Sydney Morning Herald*, 5 July 1935, p. 11; *Kalgoorlie Miner*, 5 July 1935, p. 5; *Daily Standard*, 12 July 1935, p. 4; *The Cattle King*, Angus & Robertson, Sydney, 1936; *Catholic Freeman's Journal*, 24 February 1938, p. 58.

CHAPTER 19
Blitzkrieg

For the nature columns, see *Argus*, 6 May 1939, p. 14; *Daily Mercury*, 29 May 1939, p. 4; *Townsville Daily Bulletin*, 6 June 1939, p. 5. For the scale of the irruption, see *Courier-Mail*, 20 August 1940, p. 5; *Townsville Daily Bulletin*, 2 April 1941, p. 4; 14 August 1947, p. 7. For the rats early in 1940, see *Advertiser*, 18 March 1940, p. 18; *News*, 14 March 1940, p. 10; 30 March 1940, p. 5. For their spread, damage and numbers, see *Daily Mercury*, 14 August 1940, p. 3; 6 March 1941, p. 5; 8 March 1941, p. 6; *Argus*, 30 August 1940, p. 2; 1 November 1941, p. 3; *Courier-Mail*, 14 August 1940, p. 3; 10 September 1940, p. 5; *Townsville Daily Bulletin*, 26 September 1940, p. 4; 16 October 1940, p. 4; 27 February 1941, p. 7; 21 March 1941, p. 4; 2 April 1941, p. 4; *Cloncurry Advocate*, 10 January 1941, p. 4; *Longreach Leader*, 25 January 1941, p. 12; 1 February 1941, p. 3; 15 February 1941, p. 4. For the response of children, see *Longreach Leader*, 8 February 1941, p. 20; 23 August 1941, p. 20; 18 October 1941, p. 24. For war analogies, see *Daily Mercury*, 14 February 1941, p. 4; *News*, 20 August 1941, p. 5. For rats and bushfires, see *Queensland Country Life*, 14 November 1940, p. 5; 2 October 1941, p. 3; 13 November 1941, p. 3; 4 December 1941, p. 3. For the rat in New South Wales, see *Argus*, 11 October 1941, p. 8. Crombie's paper is 'Rat Plagues in Western Queensland', *Nature*, 154, 1944, pp. 803–4.

CHAPTER 20
Taipan

For the delight at the plague's end, see *Evening Advocate*, Innisfail, 19 January 1943, p. 4. For the 1950–51 plague, see *Brisbane Telegraph*, 10 May 1951, p. 24; *Sunday Mail*, 11 February 1951, p. 3; *Longreach Leader*, 13 February 1953, p. 22. Dunnett's paper is 'Preliminary Note on a Rat Plague in Western Queensland', *Wildlife Research*, 1, 1956, pp. 131–2. James Carstairs' key papers on the rat are 'The Distribution of *Rattus villosissimus* (Waite) during Plague and Non-plague Years', *Australian Wildlife Research*, 1, 1974, pp. 95–106; 'Population Dynamics and Movements of *Rattus villosissimus* (Waite) during the 1966–69 Plague at Brunette Downs, Northern Territory', *Australian Wildlife Research*, 3, 1976, pp. 1–10. His unpublished 1971 PhD thesis, 'The Correlation of Anatomical Changes with Population Changes in *Rattus villosissimus* during the 1966–69 Plague', is also a rich source. Newsome and Corbett wrote 'Outbreaks of Rodents in Semi-arid and Arid Australia: Causes, Preventions & Evolutionary Considerations', in I. Prakash & P. K. Ghosh (eds), *Rodents in Desert Environments*, Junk, The Hague, 1975, pp. 117–153. Corbett wrote more in his book *The Dingo*, CSIRO, Melbourne, 1995. See also 'The Elusive Plague Rat', *Ecos*, 14, 1977, pp. 10–13. The South Australian Museum's work is reported in its annual report for 1969–1970. The main statement of the work of Watts and Aslin is *The Rodents of Australia*, Angus & Robertson, 1981, a key text. Serventy's account is in *The Desert Sea: The Miracle of Lake Eyre in Flood*, Macmillan, Melbourne, 1985. Rabig wrote 'Letter-winged Kite, *Elanus scriptus*, in South West Queensland', *Sunbird*, March 1970, pp. 24–6. On the inland taipan, see 'The Rediscovery of the Western Taipan', *Queensland Museum Explorer*; Jeanette Covacevich & Charles Tanner, 'Come from Nowhere...Then Just Disappear', *Australian Natural History*, 21 (1), 1983, pp. 11–13; Jeanette Covacevich, 'Two Taipans!', in *Toxic Plants and Animals: An Australian Guide*, Queensland Museum, Brisbane, 1987; Jeanette Covacevich, '*Dandarabilla* and *Gunjjiwiri*: The Discovery of the Taipains, the World's Most Dangerous Snakes', in J. Pearn (ed.), *Some Milestones of Australian Medicine*, Australian Medical Association, Brisbane, 1994.

CHAPTER 21
Small Tigers

Lawson's 'Bush Cats' first appeared in his *Short Stories in Prose and Verse*, Lawson, Sydney, 1894. On the cat's numbers and predation, see S. Legge *et al*, 'Enumerating a Continental-Scale Threat: How Many Feral Cats Are in Australia?', *Biological Conservation*, 206, 217, pp 293–303; T. S. Doherty *et al*, 'A Continental-Scale Analysis of Feral Cat Diet in Australia', *Journal of Biogeography*, 42, 2015, pp. 964–75. On the cat's consumption of the rat, see Stephanie J. S. Yip *et al*, 'Diet of the Feral Cat, *Felis catus*, in Central Australian Grassland Habitats during Population Cycles of its Principal Prey', *Mammal Research*, 60, 2015, pp. 39–50. On Ingram's brown snake, see Stephen Phillips, 'Aspects of the Distribution, Ecology and Morphology of Ingram's Brown Snake *Pseudonaja ingrami* (Elapidae)', in Daniel Lunney & Danielle Ayers (eds), *Herpetology in Australia: A Diverse Discipline*, Royal Zoological Society of NSW, Sydney, 1993. For the rat at Ethabuka, see M. Predavec & C. R. Dickman, 'Population Dynamics and Habitat Use of the Long-Haired Rat (*Rattus villosissimus*) in South-Western Queensland', *Wildlife Research*, 21, 1994, pp. 1–9. Peter McRae's thesis is 'Aspects of the Ecology of the Greater Bilby, *Macrotis lagotis*, in Queensland', University of Sydney, 2004. For Pettigrew at Davenport

Downs, see Dot Butler, 'Easter among the Bilbies', *Sydney Bushwalker*, January 1993, pp. 2–3; J. D. Pettigrew, 'A Burst of Feral Cats on the Diamantina: A Lesson for the Management of Pest Species?' in G. Siepen & C. Owens (eds), *Cat Management Workshop Proceedings*, Queensland Department of Environment, Brisbane, 1993, pp. 25–32. For better names for Australia's rodents, see Richard W. Braithwaite *et al*, *Australian Names for Australian Rodents*, Australian Nature Conservation Agency, Canberra, 1995. For the rat at the mound springs, see Janet Furler & Richard Willing (eds), *Expedition Witjira: Interim Report*, Scientific Expedition Group, Adelaide, 2006. For fossils, see Dirk Megirian *et al*, 'The Mygoora Local Fauna: A Late Quaternary Vertebrate Assemblage from Central Australia', *The Beagle: Records of the Northern Territory Museum of Arts and Sciences*, 18, 2002, pp. 77–93. For the Ord River, see D. R. King, 'The Sensitivity of *Rattus Villosissimus* to 1080', *Australian Mammalogy*, 17, 1994, pp. 123–24. For the rat on Ethabuka again, see Aaron C. Greenville *et al*, 'Extreme Rainfall Events Predict Irruptions of Rat Plagues in Central Australia', *Austral Ecology*, 38, 2013, pp. 754–64. For how Davenport Downs became Astrebla National Park, see Paul S. Sattler, *Five Million Hectares: A Conservation Memoir 1972–2008*, Mt Cotton, 2014. For the cat as a tiger, see Des Houghton, 'State Government Declares War on Feral Animals', *Courier Mail*, 20 July 2013. For the experiment at Wongalara, see A. S. K. Frank *et al*, 'Experimental Evidence that Feral Cats Cause Local Extirpation of Small Mammals in Australia's Tropical Savannas', *Journal of Applied Ecology*, 51, 2014, pp. 1486–93.

EPILOGUE

On the melomys, see John C. Z. Woinarski *et al*, 'The Contribution of Policy, Law, Management, Research, and Advocacy Failings to the Recent Extinctions of Three Australian Vertebrate Species', *Conservation Biology*, 31, 2016, pp. 13–23; Natalie L. Walker *et al*, 'The Bramble Cay Melomys *Melomys rubicola* (Rodentia: Muridae): A First Mammalian Extinction caused by Human-induced Climate Change?', *Wildlife*

Research, 44, 2017, pp. 9–21; Graham R. Fulton, 'The Bramble Cay Melomys: The First Mammalian Extinction due to Human-induced Climate Change', *Pacific Conservation Biology*, 23, 2017, pp. 1–3. On Australian extinctions more generally, see John C. Z. Woinarski *et al*, *The Action Plan for Australian Mammals 2012*, CSIRO, Melbourne, 2014; John C. Z. Woinarski *et al*, 'Australian Mammals: Have They a Future?', *Wildlife Australia*, 51, 2014, pp. 24–29. On the rat at Kununurra, see David Andrews, *The Complete Guide to Finding Australian Mammals*, CSIRO, Melbourne, 2015, p. 192. For climate change and the rat, see Aaron C. Greenville *et al*, 'Desert Mammal Populations are limited by Introduced Predators rather than Future Climate Change', *Royal Society Open Science*, 4, 2017, 170384; Chris R. Pavey *et al*, 'The Role of Refuges in the Persistence of Australian Dryland Mammals', *Biological Reviews*, 92, 2017, pp. 647–64. On the kite, see Penny Olsen, *Birds of Prey and Ground Birds of Australia*, Angus & Robertson, Sydney, 1993; Sean Dooley, 'The Final Chapter: Is the Script for the Letter-winged Kite drawing to its Close?', *Australian Birdlife*, 4(4), December 2015, pp. 34–39; Ian Cresswell & Helen Murphy, *Australia: State of the Environment 2016, Biodiversity*, Australian Government, Canberra, 2017.

ILLUSTRATIONS

1. Angus Emmott, *Long-haired Rat*, 2011.
2. H.C. Richter, *Mus longipilis*, 1854, from John Gould's *Mammals of Australia*.
3. Ludwig Becker, *Border of the Mud-Desert near Desolation Camp*, 1861, State Library of Victoria.
4. Ludwig Becker, *Long-haired Rat*, 1861, State Library of Victoria.
5. Unknown photographer, *Alfred North, Kenric Harold Bennett and Edward Pierson Ramsay in the Boardroom of the Australian Museum*, State Library of New South Wales.
6. Alfred Pearse, *Louis de Rougemont and his dog Bruno find Safety up a Tree from the Long-haired Rat*, from *Wide World Magazine*, February 1899.
7. Sidney Jackson, *Young Letter-Winged Kites at Camp with Daily Meals of Rats*, from *Emu*, January 1919.
8. H.H. Finlayson, *Jimmy Naylon Arpilindiḵa*, 1931, South Australian Museum.
9. H.H. Finlayson, *The Oolacunta As He Appears At Speed*, 1931, from *The Red Centre: Man and Beast in the Heart of Australia*, 1935.
10. Angus Emmott, *A Night's Catch of Long-haired Rats*, 2011.

INDEX